SpringerBriefs in Environmental Science

SpringerBriefs in Environmental Science present concise summaries of cutting-edge research and practical applications across a wide spectrum of environmental fields, with fast turnaround time to publication. Featuring compact volumes of 50 to 125 pages, the series covers a range of content from professional to academic. Monographs of new material are considered for the SpringerBriefs in Environmental Science series.

Typical topics might include: a timely report of state-of-the-art analytical techniques, a bridge between new research results, as published in journal articles and a contextual literature review, a snapshot of a hot or emerging topic, an in-depth case study or technical example, a presentation of core concepts that students must understand in order to make independent contributions, best practices or protocols to be followed, a series of short case studies/debates highlighting a specific angle.

SpringerBriefs in Environmental Science allow authors to present their ideas and readers to absorb them with minimal time investment. Both solicited and unsolicited manuscripts are considered for publication.

More information about this series at http://www.springer.com/series/8868

Frederic R. Siegel

Adaptations of Coastal Cities to Global Warming, Sea Level Rise, Climate Change and Endemic Hazards

 Springer

Frederic R. Siegel
George Washington University
Washington, DC, USA

ISSN 2191-5547   ISSN 2191-5555   (electronic)
SpringerBriefs in Environmental Science
ISBN 978-3-030-22668-8   ISBN 978-3-030-22669-5   (eBook)
https://doi.org/10.1007/978-3-030-22669-5

This Springer imprint is published by the registered company Springer Nature Switzerland AG
The registered company address is: Gewerbestrasse 11, 6330 Cham, Switzerland

I dedicate this book to those that plan for the contemporary and future security of populations and assets in coastal cities (and those inland) without over-influence from political-economic interests but with the foremost interests being for all peoples in today's societies, their children and grandchildren, and ecosystems that support them.

Prologue

There were at least 166 seaport cities with populations greater than one million people in 2018. Most are in Asia with 35 in China, 8 in Japan, 7 in India, and 26 in the rest of Asia. Many coastal cities with one million plus populations are in Africa and the Latin America-Caribbean region, each with 23 cities [1, Appendix]. This compares with at least 136 such cities reported in 2005 [2]. The increase in the number of these high population coastal cities is the result of the growth in global maritime commerce for developing nations and demographic changes from natural growth but mainly from rural citizens moving to coastal (and inland) cities for employment, better access to health services, and improved education for children. The number of seaports in Europe, North America, the Middle East, and Oceania has shown little increase.

Of these 166 coastal cities, there are 23 of the world's 37 mega-cities with populations ranging from more than 10 million inhabitants in Nagoya, Japan, to more than 38 million in Tokyo. Their aggregate population is more than 400 million people and when added to the other coastal cities with at least one million inhabitants, the total population exceeds 800 million people or about 10% of the Earth's population [1]. The highly and densely populated port cities and other less populated ones are at increased risk of future flooding, extreme weather events (e.g., storms, drought, heat), availability of safe water, concerns about food security, and potential for contracting an infectious disease, all intensified in great part by global warming/climate change.

The rise in the earth's temperature from pre-industrial time to the present is ~1 °C. This contributes to the global problems cited in the previous paragraph. Projections on increases to sea level rise and climate changes from global warming for a global mean surface temperature rise to 1.5 °C with conditions further projected for 2.0 °C and ramifications for coastal cities were published in a June, 2018 Royal Society paper [3]. They were further elaborated on in an October, 2018 Intergovernmental Panel on Climate Change report. Within the report, one chapter deals with implications for coastal areas including small islands (and island states), deltas and estuaries, and wetlands [4].

Port cities may also be at risk of and vulnerable to natural and anthropogenic hazards (e.g., earthquakes, exposure to pollution) unrelated to or indirectly related to global warming/climate change. It is clear that the potential seaport city problems noted above are interacting and that adaptations to mitigate or eliminate them have to consider the interactions. This book highlights coastal city problems and suggests adaptations that can help resolve them now and for future generations.

References

1. Demographia (2018) World urban areas (Built up urban areas or world agglomerations), 14th annual edition, 118 p. Seaport designations were reviewed from internet searches, www.demographia.com/db-worldua.pdf
2. Hanson S, Nicholls R, Ranger N, Hallegate S, Corfee-Morlot J, et al (2011) Climate change, Dordrecht 104:89–111
3. Nicholls RI, Brown S, et al (2018) Stabilization of global temperatures at 1.5°C and 2.0°C: implications for coastal areas. Philos Trans R Soc A Math Phys Eng Sci 376(2119):20160448. https://doi.org/10.1098/rsta.2016.0448
4. Intergovernmental Panel on Climate Change (2018) Global warming of 1.5°C. Chapter 3. Impacts of 1.5°C of global warming on natural and human systems. United Nations, Geneva, pp 175–311. https://report.ipcc.ch/sr15/pdf/sr15_spm_final.pdf

Contents

Chapter 1
Introduction

There are four choices that coastal cities have as threats from global warming, climate change, and population growth put their citizens at risk. The risks are of injury or death, loss, or diminished availability of life-sustaining natural resources, loss of citizens' property and commercial-industrial operations vulnerable to damage and destruction and employment they offer, and disruption or destruction of cities' infrastructure. The viable choices are to build defenses, opt for a managed retreat, and adaptation to mitigate or eliminate risk and vulnerability. A fourth choice would be to maintain the status quo but this is not realistically viable either politically, sociologically, or economically.

1.1 Build Defenses

The first of these is to build hard defenses against dangers from the ocean (e.g., sea level rise, storm surges, and even tsunamis). These include dikes, breakwaters, and seawalls that will be discussed in Chap. 3. Defenses such as retention dams and spillways also have to be in place against floods that originate from storms that move inland onto a drainage basin and can cause flooding in downflow coastal urban centers. Both types of defenses have to be monitored and maintained. Soft defenses are nature based and include tidal marshes, mangrove vegetation (forests), and sand dunes that should be preserved or restored if necessary. These damp wave energy and shore erosion also reduce the energy of storms that track across them in their path inland. Socio-economic defenses against extreme weather conditions such as long-term droughts or heat waves will protect citizens, water and food security, and their health. Hard and soft defenses can be put in place to protect coastal cities from high energy wind-driven extreme storms (hurricanes, typhoons, monsoons) that seasonally impact highly and densely populated coastal cities bordering

the Atlantic, Pacific, and Indian oceans. These defenses can range from economically and technically doable in some cities/countries but not in others without assistance from international banking institutions and help from countries with strong
monetary reserves and technical expertise.

1.2   Managed Retreat

A second choice is to implement a managed retreat from global warming/climate
change dangers, but this is an option that is not economically feasible for some
coastal cities, especially in less developed and developing countries. For example,
Jakarta, Indonesia is a coastal mega-city (>32 million inhabitants) that sits in a lowland sloping 0–2°. The city subsided 3–10 cm (~1–4 in.) annually from 1974 to
2010 as a result of overuse of aquifers as the main water source. In some areas of
Jakarta, the rate is now 20–25 cm (8–10 in.) annually due in part to increasing
urbanization and draw on the aquifer water. Some developed coastal areas are up to
2 m (~6 ft) below sea level and suffer regular tidal and storm surge flooding. The
Indonesian government calculated that it would cost at least US$220 billion to move
the north part of Jakarta to the elevated southeast part of the city that is 50 m (164 ft)
above mean sea level [1]. The government opted to use technical assistance from
Dutch experts to replace a failing, subsiding 40-year-old seawall with a 15 mile
(>25 km) long, massive seawall across Jakarta Bay at a cost of US$32–40 billion
[2]. A major problem that has to be solved is the chemical and biological pollution
now in Jakarta Bay that will intensify in the lagoon setting unless waste treatment
plants are built to clean wastewater from rivers and canals that discharge into the
bay [3]. Funding for the project is from the national government, the municipality
of Jakarta and is supplemented by real estate investment. When completed, post
2025, it would create a large lagoon with reclaimed land that is planned to be com-
mercialized. What may be a like option that would require great investment is not
just to effect a managed retreat of part of an at-risk population and a city's economic
force inland but to build a new city with all defenses needed for future protection.

1.3   Adaptation

Biologically, adaptation is an inbred organism characteristic that promises survival
and species procreation. It can be something produced that helps sustain life by
adjusting to different situations or uses such as changes in environmental conditions. For human (e.g., coastal) populations, this means building an adaptive capacity. Adaptation has multiple facets to achieve that capacity. An adaptation may be
physical, (e.g., a sea dike), chemical (e.g., pollution control), biological (e.g., vaccination), cultural (e.g., adjust traditional practices), behavioral (e.g., conserve
water), structural (e.g., coordination by adaptation evaluation teams),

developmental (e.g., environmental education), technological (e.g., use best technology available or affordable), or combinations of these as changing conditions dictate.

Changing is a keyword because adaptation during change (e.g., in coastal areas) is likely easier done and at better benefit/cost economics than if we wait for a change to reach a stage of meta-stability. For humans, adaptation means becoming better suited to a natural environment. Ideally, any adaptation(s) selected to bring populations into balance with changing environments should be sustainable (e.g., a seawall). Flexible (e.g., able to be heightened), based on fact or best available data and projections (e.g., of sea level rise), and prioritized (e.g., sited where it will be most effective and equitably beneficial to all citizens regardless of economic status).

In some cases, adaptations may be designed to undo negative results of former human adaptation to an environment. This may be by restoring wetlands and planting mangrove vegetation that have been drained or cut away in the past to open space for human use. It may be by heightening sand dunes and planting them with native grasses for stabilization and to trap sand. Working with nature to achieve positive adaptation is a class of managed retreat [4]. As noted above, restoration of wetlands and mangrove forests can help protect coastal areas by damping wave energy and the effects of hurricanes (typhoons, monsoons) as they track landward across these shore environments.

References

1. Abidin HZ, Andreas H, Gumilar J, Brinkmann JJ (2015) Study on the risks and impacts of land subsidence in Jakarta. Proc Int Assoc Hydrol Sci 372:115–120. https://doi.org/10.5104/pihas-372-115-2015
2. Win TL (2017) In flood-prone Jakarta, will Giant Sea Wall plan sink or swim? Reuters
3. Breckwoldt A, Dsikowitzky L, Baum G, Ferse SCA, van der Wulp S, Kusumanti I, Ramadhan A, Adrianto L (2016) A review of stressors, uses, and management perspectives for the larger Jakarta Area, Indonesia. Mar Pollut Bull 110:790–794. https://www.researchgate.net/publication/307157229
4. Dept. for Environment, Food and Rural Affairs (2010) Natural environment: adapting to climate change. London, 51 p. http://www.dedra.gov.uk/environment/climate/index.htm

Chapter 2
Arresting/Controlling Saltwater Contamination of Coastal Aquifers

2.1 Root of the Problem and Some Pre-planning

The demand for water in coastal zones increases with expanding urbanization. As populations grow, their needs for potable water, safe water for cooking and personal hygiene grows as does water for food security (e.g., for irrigation agriculture, animal husbandry). Added to this is the desire of many governments to attract manufacturing/industrial development that requires much water but creates employment and sources of taxation. The cheapest source of water is from aquifers because groundwater requires less treatment than would surface waters that may be available. Cities worldwide are growing so that in 2018, 4.1 billion people of the world population of 7.6 billion live in cities and urban agglomerations. One billion live in coastal regions. City populations are projected to increase to 6.9 billion people of the 9.9 billion global population in 2050 [1]. The number of people living in coastal cities is likely to increase as well. There are problems today in meeting the safe water requirements of people, agriculture, and manufacturing/industrial projects that include water from coastal aquifers, so what can be expected in the future?

Planning to tap a coastal aquifer should draw on what past experience has shown. First, for a reliable and continuous source of freshwater (given no overuse), an aquifer should be unconfined, thus rechargeable by rainwater or snow/ice melt that seeps into it through an overlying soil/rock cover or from stream/river inflow. Next, a well should be spudded in more than 50 m (~163 ft) inland, a number that will increase with rising sea level. Siting where there is a steeper slope inland is preferred to a gentle or moderate slope because a steeper slope would produce a stronger hydraulic gradient for groundwater moving seaward thus keeping seawater at bay. Plans should avoid drilling excessively deep because this can contribute to a lessening of hydrostatic gradients where seawater, heavier than freshwater, could form a wedge that would otherwise invade the bottom of a coastal aquifer. A high rate of pumping from a multi-well system should be prohibited because the water table level will

F. R. Siegel, *Adaptations of Coastal Cities to Global Warming, Sea Level Rise, Climate Change and Endemic Hazards*, SpringerBriefs in Environmental Science, https://doi.org/10.1007/978-3-030-22669-5_2

drop with a corresponding drop in hydrostatic pressure since recharge cannot keep up with discharge. There should be no practice of hydrofracturing within 100 m (~327 ft) of a coast as a minimum.

2.2   Hydrostatic Principle

Modern models dealing with seawater intrusion of coastal aquifers are based on the principle presented by Ghyben (1888) and Herzberg (1901) [2, 3]. The principle is represented by the equation

$$z = \rho_f / (\rho_s - \rho_f) \cdot h$$

Figure 2.1 illustrates the Ghyben–Herzberg relation. In the equation, the thickness of the freshwater zone above sea level is represented as h and that below sea level is represented as z. The two thicknesses h and z are related by ρ_f and ρ_s, where ρ_f is the density of freshwater and ρ_s is the density of seawater. Freshwater has a density of about 1.000 g/cm^3 at 20 °C, whereas that of seawater is about 1.025 g/cm^3. The equation can be simplified and $Z = 40H$.

Therefore, the depth of the seawater interface below mean sea level is 40× the elevation of the water table above sea level. Thus, if a water table in an unconfined coastal aquifer is lowered by 1 m, the freshwater–seawater interface will rise by 40 m. The principle is fine but represents a steady-state condition and does not account for other characteristics of a coastal aquifer that affect discharge/recharge kinetics.

As clear from many published studies, modeling presents a complex problem because each aquifer has its unique characteristics that have to be determined before any calculation is attempted. In planning to protect coastal aquifers from global warming, estimates have to be input to models that reflect projected sea level rise

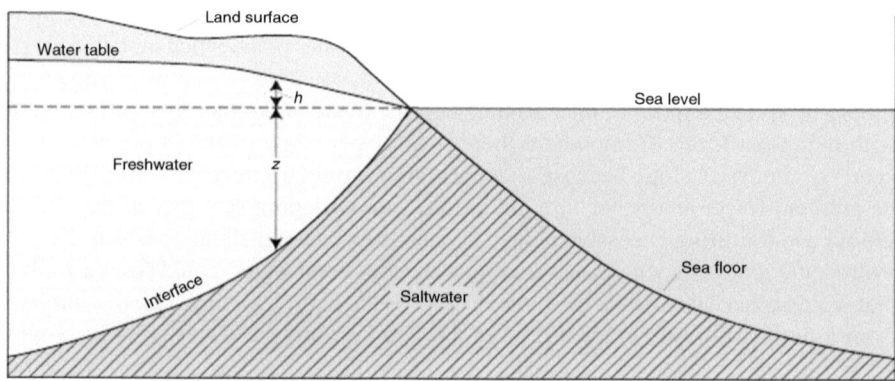

Fig. 2.1 A graphic representation of the Ghyben–Herzberg relation [4]

[5]. This includes the hydrostatic gradient set by the difference in height between sea level and the highest level of the water in the inland section of an aquifer. Add to this the depth and thickness of the aquifer, its porosity (ability to hold water) and permeability (capacity to transmit water), the area underlain by the aquifer, its recharge areas (upstream hydrologic conditions), the presence or not of fractures in the bounding rock units, tidal cycles, and seasonal fluctuations (discharge from the aquifer). Apart from modeling and an aquifer's relation to sea level rise is reality when there is a continuous and long-term draw on the aquifer during times of extended and severe drought. In Monterey County, California, USA, agricultural overuse of aquifer waters during years of severe drought (December, 2011 to March, 2017) and hence lessening of the hydrostatic gradient allowed the seawater boundary to reach more than 10 km (6 mi) inland. In the same area, the salt water intrusion was halted by stopping pumpage from the aquifer because of an investment in two treatment plants that recycled wastewater and distributed the safe water through 32 km (20 mi) of a pipe network to growers along the California coast [6].

It should be noted that there is an apparent increase in sea level and a real increase of hydrostatic gradient that favors seawater intrusion of coastal aquifers where there is subsidence because of over discharge of groundwater not balanced by aquifer recharge. This leads to added jeopardy for populations and urban assets because of seawater encroachment of land and greater risk exposure to storm surges and flooding. Examples of this are Ho Chi Minh City, Bangkok, and Jakarta and will be discussed in Chap. 5 [7].

2.3 Pressure Pushing Seaward Vs. Pressure Pushing Inland

Coastal freshwater aquifers worldwide are in contact with seawater where they outcrop at the continental slope. These aquifers service human water needs and sustain agricultural and industrial projects. The freshwater–seawater contact zone is in equilibrium or favors groundwater when the pressure of freshwater in the aquifer discharging to the sea (hydrostatic gradient) equals or exceeds the pressure of the seawater driving to access the aquifer. However, when there is a continuous heavy draw on the groundwater, the aquifer pressure drops and the hydrostatic gradient may not be able to maintain equilibrium so that the denser seawater moves into the aquifer as a bottom wedge (toe). When this wedge moves inland, salt water contaminates aquifer water. Humans can tolerate a slight degree of salinity in their water supply as can some agricultural crops and food animals. However, with increases in salinity, water becomes less potable and may not serve farm or industrial needs. The US Environmental Protection Agency advises that water with salinity <80 mg/L is excellent for drinking whereas that with 80–500 mg/L is considered good, 500–800 mg/L is fair, 800–1000 mg/L is poor, and groundwater with >1000 mg/L is not considered potable [8].

Rising sea level caused by global warming (e.g., from seawater thermal expansion and melting of continental and alpine glaciers and ice sheets) favors salt water intrusion into coastal aquifers by overcoming the hydrostatic gradient exerted by the freshwater aquifer. This pressure will increase as the rise in sea level increases. The Intergovernmental Panel on Climate Change projects that by the end of the century the rise will be in the range of ~11–54 in. (~28–137 cm) [9]. One model calculation method suggested that a sea level rise of 20 in. (~50 cm) would push a saline wedge landward to up to 50 m. Another model calculation suggested that the landward intrusion could be more than a kilometer [10, 11]. Still another model predicted that a 1 m (39.3+ in.) sea level rise would move a seawater wedge 15–30 m (~50–100 ft) towards water extraction wells.

2.4 Methods Used to Prevent Salt Water Intrusion

There are several methods that may effectively reduce the possibility of salt water intrusion of coastal freshwater aquifers [5]. Each is related to maintaining or increasing the hydrostatic gradient active in the aquifer or reducing access of seawater into the aquifer from surface sources. The latter includes disallowing either excavating navigation channels, or agricultural or drainage channels. These can drop the aquifer level thus reducing hydrostatic pressure that leads to intrusion. Users of groundwater from coastal aquifers want to be able to extract their freshwater needs without causing a lessening of hydrostatic pressure in the aquifer that leads to salt water intrusion. This may not always the case because as noted in the previous paragraph, there can be high volume groundwater discharge during drought that is not balanced by recharge with a consequent drop in the aquifer hydrostatic pressure. In the long term, hydrostatic pressure in an aquifer in a non-drought coastal or inland region may drop because of urbanization that covers recharge areas with housing and infrastructure, or because of diversion of streams/rivers that inflow water to aquifers. Municipalities can set conservation norms for users to reduce extraction to maintain hydrostatic pressure by cutting down the draw on aquifer resources. When there has been drought that affects groundwater levels, communities have treated and recycled wastewater and expanded recycling systems in order to reduce extraction of groundwater. Communities may opt to inject water into an aquifer to stabilize its hydrostatic gradient [2, 3]. Users also seek alternate sources for their water needs. Recharge has been enhanced by ponding surface water and storm water runoff (artificial infiltration) or redirecting river water temporally to recharge areas.

Barriers to seawater intrusion of coastal aquifers include protective troughs developed in a string of wells parallel to a coast that extract seawater behind a seawater wedge thus reducing pressure landward and halting the advance of intruding salt water. Pressure ridges using a string of freshwater injection wells parallel to a coast have also been used to maintain a hydrostatic pressure in an aquifer that keeps out seawater. Relatively shallow coastal aquifers have been protected by strings of wells also parallel to a coast that inject gel cement (grout) or insert steel plates into an aquifer front to create a physical barrier to seawater intrusion [5].

References

1. World Population Data Sheets (2018) Population Reference Bureau, Washington, DC. www.prb.org/2018-world-population-data-sheet
2. Ghyben WR (1888) Nota in verband met de voorgenomen putboring nabij Amsterdam. Tijdschrift van Let Koninklijk Instituut van Ingenieurs, The Hague, Netherlands, pp 8–22
3. Herzberg A (1901) Die Wasserversongung einiger Nordseebader. Wasserversongung 44:815–819, 842–844
4. Barlow PM (2003) Ground water in a freshwater-saltwater interface in a coastal water-table aquifer. US Geological Survey Circular, Washington, DC, p 1262. (Unpaginated)
5. Kumar CP (2016) Sea water intrusion in coastal aquifers. EPRA Int J Res Dev 1:27–31
6. Walton B (2015) Here comes the sea: the struggle to keep the ocean out of California's coastal aquifers. Water News, Circle of Blue. (Unpaginated) https://www.circleofblue.org/2015/world/here-comes-the-sea
7. Siegel FR (2018) Cities and mega-cities: problems and solution strategies. Springer Briefs in Geography, 117 p
8. EPA (1973) Identification and control of pollution from salt water intrusion. EPA-430/9-73-013, Washington, DC, 104 p
9. Intergovernmental Panel on Climate Change (2013) Fifth assessment report. In: Climate change 2013. The physical science basis. Summary for policy makers. Cambridge Univ. Press, Cambridge, 29 p
10. Werner AD, Simmons CT (2009) Impact of sea-level rise on sea water intrusion in coastal aquifers. Groundwater 47:197–204
11. Ketbchi H, Mahmoodzadeh D, Ataie-Ashtiani B, Simmons CT (2016) Sea-level rise impacts on seawater intrusion in coastal aquifers: review and integration. J Hydrol 535:235–255

Chapter 3
Structures That Protect Coastal Populations, Assets, and GDPs: Sea Dikes, Breakwaters, Seawalls

In 2018, about one billion people of the Earth's 7.6 billion lived in marine coastal zones. The people, their property, and the infrastructure that supports them, and a city or national per capita GDP are at risk at multiple levels. These include coastal erosion, high and spring tides that cause lowland flooding, weather-related events (storm surges, flooding, wind, crop loss, unprotected anchorage, stabilization of navigation channels, and rarely, killer tsunamis). These coastal zones now, and more so in the future, are likely to be at high risk because of global warming-driven sea level rise. Human activity inshore can increase the level of risk from the above cited sources, such as flooding by abetting subsidence because of overuse of coastal aquifers for a water supply. Dikes, breakwaters, seawall, and related structures are designed to thwart for some time (50 years?) damaging, destructive forces that assault coastal regions worldwide. They are costly to build and maintain but in short and long terms present economic benefits that preserve much, much more in capital investment.

3.1   Sea Dikes

Dikes are human-built broad-based and elevated earthen structures that protect low-lying coastlines from flooding and erosion and support coastal protection defenses where sand dunes are present. They also serve to prevent encroachment and loss of terrain to sea level rise.

Dutch engineers, marine geologists, computer modelers, and other experts team to design and build sea dikes tailored to different low-lying coastal environments. They have successfully protected the Netherlands from flooding and erosion by reducing North Sea wave impact and run up (e.g., from overtopping during spring tides, storms).

© The Author(s), under exclusive license to Springer Nature Switzerland AG 2020   11
F. R. Siegel, *Adaptations of Coastal Cities to Global Warming, Sea Level Rise, Climate Change and Endemic Hazards*, SpringerBriefs in Environmental Science, https://doi.org/10.1007/978-3-030-22669-5_3

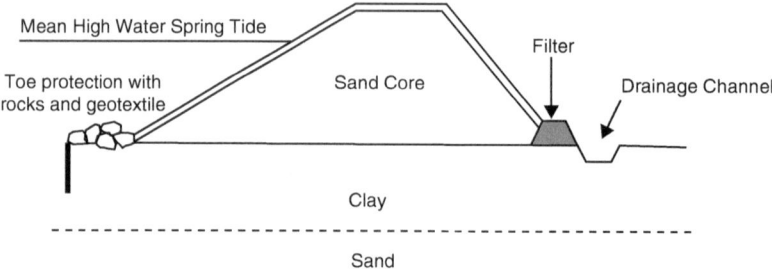

Fig. 3.1 Typical sea dike cross section [1]

The Dutch experience in sea dike construction follows guidelines that adapt to coastal topography and offshore hydrography plus the historical record of events that have impacted a coastal region designated for dike emplacement. Construction specifications would include the following as illustrated in Fig. 3.1 [1].

(a) **Reduce wave action**; seaward facing slope should have a gradient between 1:3 and 1:6
(b) **Maximize stability**; the inland facing slope should have a gradient of 1:2 and1:3
(c) A **preferred core earthen material is sand**; this assures that water that enters can drain into a (gravel) channel back to the sea, thus preventing weakening of the core by water saturation
(d) **Protect the sand core**; the dike should have an impermeable cover layer (armor) that sometimes may be clay covered by asphalt, concrete slabs, stones, or even grass; a steeper slope (see (a) above) would require additional armoring to withstand more direct wave impact
(e) **Prevent scouring or undercutting (undermining)**; the dike should have an armored (rocks, rubble) base or toe
(f) **Dike design**; specifications should also include **adaptation** plans to raise dike height as necessary **to counter sea level rise** with an increased height and broader base (need more space inland). Land will have to be given up for existing or planned uses in order to support coastal protection

3.1.1 Costs of Sea Dikes

Sea dikes are costly to build and prices vary worldwide. Costs incurred are dependent on the country, lost land use where dikes are to be constructed, cost of property acquisition, labor and materials costs, plus, infrastructure that has to be relocated, dike design and modeling, and project management including regular inspections and maintenance [2]. To raise a dike for **urban** protection in the Netherlands 1 km long with a 1 m height could range in cost in 2012 k-euros from k-€15,900 to k-€23,000. In Canada, the 2012 figure for urban protection could range from k-€1600 to k-€12,400. To protect a **rural** environment could cost from k-€1000 to k-€3300 in Canada, k-€4700 to k-€11,500 in the Netherlands, and k-€700 to k-€1200 in Vietnam. Maintenance costs vary as well so that in Vietnam they would range from €24,000 per linear kilometer annually whereas in the Netherlands the average cost is €108,000 per linear kilometer [2].

3.1.2 A New Type of Sea Dike Being Used

Hillblock is a relatively new type of revetment made of concrete blocks with a unique shape (Fig. 3.2) that is stable under relatively severe wave loads and designed to reduce the volume of water flowing over a dike due in part to its cast concrete surface infiltration sites. The individual pillars are cast in half sections and placed together without cement. They are tailored to each project and are 20–50 cm (~7.9–19.6 in.) high. With a base side of 25 × 25 cm (~10 × 10 in.), they are pinched in the center of their height like a weight dumbbell. The 20 cm (~7.9 in.) high block weighs 10 kg (22 lb) and the 50 cm (~19.7 in.) high block weighs 23.5 kg (~51.7 lb). Block surfaces have beveled edges and beveled corners that present a series of surface infiltration openings through which seawater run up seeps into channels below and discharge into drainage channels that carry the infiltrated water back to the sea. This reduces the volume of seawater that may flow over the dike. The blocks are set in a 15 cm granular (gravel) layer finished smooth and level and underlain by a layer of geotextile that directs drainage and protects the underlying earth from being washed away. Once the blocks are set in place, the space between the bases are filled with dry grout that keeps the blocks in place without the possibility of being pulled out by water pressures during wave impact. Placed against one another the blocks act as a single unit [3]. Tests were run at the Deltares new Delta Flume that can simulate

Fig. 3.2 A Hillblock pillar in profile and Stolford, UK revetment (see following paragraph)

wave action up to 5 m. Results showed no slippage or movement of the units suggesting that a Hillblock revetment would be stable when relatively severe wave loads higher than 3 m impact, a force expected from a major North Sea storm [4].

Commercial and tangible benefits claimed by the producers are that the mass of concrete used in a Hillblock sea dike is 30–40 % lighter than the concrete in a traditional covered dike thus reducing usage (cost) for concrete. The design suggests a 30–50 % wave impact abatement thus reducing erosion potential and giving the possibility that sea dike seaward slopes could be lowered [5]. Hillblock sea dikes (also called block revetments) are the default revetments being built now along the Netherlands coast to mitigate varying tidal and wave/storm conditions [6]. An emplacement of at the first Hillblock revetment outside the Netherlands was dedicated during March, 2019 near the hamlet of Stolford, UK, overlooking Bristol Channel, to prevent erosion and flooding from the run up and overtopping of a protective ridge that had to be replenished annually at a cost of £50,000 (US$65,000). In past years, flooding from high tides or high tides and storms affected 24 residences, 660 hectares (1650 acres) and roads. The new revetment with blocks held in place by steel piles and concrete kerbs (curbs) is designed to protect 20 residences, 70 hectares (175 acres) of productive farmland, and roads by reducing wave impact and run up overtopping by about 30%. The Stolford revetment is 180 m (~590 ft) long and has an upslope width of about 24 m (~79 ft) for an area of ~4320 m^2 (46,500 ft^2) (**Personal Communication**, Jaap Flikweert, Leading Professional Flood Resilience, April, 2019). The cost of the project was £1.5 million (US$1.95 million).

3.2   Breakwaters

Breakwaters are sturdy offshore detached wall-like mounded structures anchored to the sea floor individually or in series. They are built dominantly with a core of small size rock overlain by a layer or layers of sequentially medium and larger size rocks (riprap) armored with large (>1 m) rock masses (riprap) of angular, crystalline rock (e.g., granite, basalt, quartzite) put in place randomly [7, 8]. Breakwaters can also be constructed of pre-cast steel reinforced concrete or built up as artificial reef barriers with trashed materials deposited on the sea floor.

Breakwater size is important as evidenced by ancient breakwaters that if once submerged have been reshaped by centuries of storms, or if once emerged, have been eroded below sea level. Both actions affected the ability to protect shorelines from changing configuration over time [9]. These rock masses are most often anchored on a slightly sloping sea floor in relatively shallow waters close to shore (e.g., 30–100 m [~99–330 ft]). Breakwater orientation will be selected by coastal engineers according to changes in shoreline configuration, sea floor topography, and dominant wind and hence incoming wave direction with the purpose of maintaining or building up beach morphology [10]. Basically, breakwaters are barriers, submerged or slightly above still sea level, that damp wave energy and break up wave advance. Some may have caps to prevent overtopping (Fig. 3.3) [11]. They

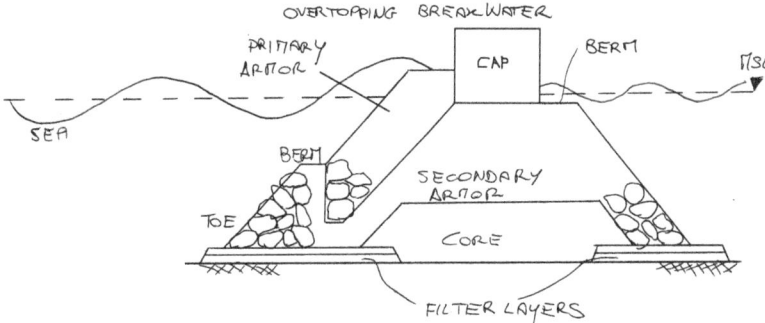

Fig. 3.3 Example of a crested non-overtopped breakwater [11]

attenuate, (disperse, absorb) wave energy as the waters impact the voids and irregular surfaces of rock masses that are porous and may be permeable (able to pass water through). Wave energy diminishes as it impacts and passes through the different layers of breaker rocks thus greatly reducing the force of incoming waves in leeward waters and minimizing coastal erosion and long shore drift. As noted above, this protects, or in some designs, builds up sand beaches (width and thickness) hence preserving the tourist incomes they generate. Breakwaters may, in some cases be attached to land at one end and curved (arcuate) in order to create or protect an anchorage (safe harbors) for ships from weather (high winds and strong wave impact) [12]. Unlike sea dikes, breakwater are not designed to counter encroachment, flooding at coastlines as sea level rises, or storms surge but are emplaced to reduce wave activity landward to protect beaches from erosion and at the same time minimize down drift erosion.

3.2.1 Construction Determinants and Norms

Before constructing the fixed (vs. floating—next paragraph) breakwaters, there are several factors that must be evaluated. First is a hydrographic survey that shows the topography of the sea floor that is used to evaluate proposed sites for individual or a series of detached structures for a given shore. The character of the sea floor with respect to the bearing strength and solidity where structures may be emplaced is essential information as is a knowledge of wave lengths and wave height during years' seasons. Together, these measurements determine the breakwater footprint (shape, length, and width [cross section]). They in turn affect whether a submerged or emerged above wave level of a breakwater or of a breakwater series is required. For a breakwater series, the measurements affect spacing between breakwaters (e.g., 50–300 m [~165–990 ft]). With these data, marine coastal engineers can calculate the amount of rock mass (or perhaps pre-cast steel reinforced concrete) that will be needed. This allows the cost of rock and its transport to the structure site to be determined, added to which is the labor cost to set rock masses in place. There is an ongoing cost of monitoring and periodically inspecting breakwaters and to maintain and repair them as needed. Breakwaters as other coastal structure defenses are expected to be effective for at least 50 years.

In some areas, floating breakwaters as pontoon or boxes are used to tamp incoming wave energy but only where a wave height is less than ~2 m (6.5 ft). They are useful where sea floor sediments cannot easily support fixed breakwaters without substantial expenditure. Floating breakwaters can be detached and stored when not needed but are costly to reintroduce to stem renewed strong wave activity [13]. Research shows that they have best results when a floating breakwater is about half as wide as an incoming wave length and/or when their natural oscillation period is much longer than the wave period [14].

3.2.2   Adaptation to Changes from Global Warming

Climate change as a result of global warming of the oceans has to be considered in breakwater future stability design. This is because of sea level rise and the likely increase in wind speeds during storms and the increase in wave heights and energy that higher velocity winds will generate [15]. Do we build for the present with a plan to raise breakwater heights in the future as warming increases with the above noted consequences? Or do we build a breakwater to counter projected, but not certain to be accurate Intergovernmental Panel on Climate Change projections for 2050? Or do we look beyond for guidance for breakwater construction parameters at more recent reports that give higher projections for sea level rise (discussed in Chap. 5). The same questions exist for the previously discussed sea dikes and next to be considered seawalls.

Experimental research on incident wave heights and breakwater properties continues in laboratory wave flumes on breakwater configurations at different distances from a coast to determine their efficiency on protecting sandy, gently sloping coastline beaches. They may confirm observations and measurements made by earlier studies but under different beach conditions. For example, in nearshore environments submerged breakwaters cause crashing waves and sediment transport similar to activity at beaches not protected by breakwaters [16].

3.3   Seawalls

Seawalls are vertical or close to vertical structures constructed at the coast and parallel to it to prevent or mitigate erosion, flooding from wind driven high/spring tides, and storm surges in a high energy dynamic environment. They may be in place to protect a family beach residence but for this book's purpose they are built to protect large areas of urban populations, their assets, municipal infrastructure, and places of employment. In specific locations, they serve as defense against a tsunami by damping its height, reducing its velocity and hence run up inshore based on where and with what characteristics past events occurred. Seawalls are often massive steel reinforced concrete structures at the land–water interface and as noted, are parallel to a stretch of coastline as protective barriers to provide public safety for onshore urban centers, to shield physical property, to secure continued functioning of infrastructure, as well as sheltering ecosystems. Their seaward bases may be armored with rubble (riprap, irregular sized, angular, blocks of quarry rock) to protect against scouring and undermining (and weakening) of the structure. Seawalls may also be constructed of rubble mounds, brick or block, or gabions (wired containers filled with rubble). Their purpose is to block wave or storm surge intensity. As such, they may be stepped to dissipate force of waters rushing shoreward. They may have a concave face seaward so that a mass of water falls back on itself muting wave or storm surge force as well as to limit flow over the wall. Whether vertical, stepped or curved concrete seawalls or walls of massive rock barriers are selected for a coastal city depends on an assessment of a location's history with respect to storms and surges, wave heights, and velocities. Management of existing structures has to consider adaptation options as sea level rises and stronger wind-driven storm events increase in frequency, intensity, and duration. Likewise, proposed walls in a planning stage or actually beginning construction have to account for projected future threats from the oceans as global warming continues with sea level rising and extreme weather events recurring [17].

Seawalls are investments built to protect life, property, and ecosystems. They are most costly when protecting urban centers, tourist attractions (beaches, resorts), critical infrastructure, and investment zones.

The Japanese government, still reeling from the Fukushima disaster in 2011, reacted by building 245 miles (~392 km) of seawalls, some as high as 41 ft (~12.4 m) and with breakwater frontage, The cost was US$12.74 billion (1.35 billion yen) or more than an average of US$53 million a mile (~1.6 km) [18]. New York City suffered the onslaught of loss of life, flooding, and damage from Hurricane Sandy in 2012. The cost to the city, individuals, and businesses was $19 billion. As a result, New York City officials reviewed plans and costs for a seawall ($20 billion) to protect Manhattan from flooding. However, officials favored an alternative plan where the city would put a series of levees in place, a flood wall, and a park that would protect the city from major flooding at a projected cost of more than $1 billion [19]. The city of Boston, USA, missed the onslaught of Hurricane Sandy by an hour and a half before high tide. The Boston City Council and University of Massachusetts teams are reviewing multiple plans for protecting the city from damaging, destructive flooding. These include a seawall across the inner Boston harbor at a projected cost of more than $2.5 billion built with a maritime opening that can be shut under the threat of strong storm activity at high tide [20]. The Thailand government reviewed the possibility of setting a seawall across the Gulf of Thailand but thought that the nation could not afford the estimated US$14 billion to build it. Given continuing sea level rise and increases in storm activity, plus the Thai problem of subsidence in Bangkok, the important Thai revenue producing capital and coastal city, this is a great mistake. Any calculation of the cost of a seawall has to include not only the initial construction price but also the long-term cost of maintaining and repairing it. As noted earlier, the minimum projected life for a seawall or other coastal defense structures is estimated at 50 years.

3.3.1   Complimentary Barriers to Seawalls

An Australian adaptation to a seawall coastal defense is to establish vegetation directly in front of them such as mangroves and sea grasses or to create a shellfish reef or an artificial reef there. Also vegetation landward of a seawall should be preserved or established if that possibility exists [21]. Experience shows that areas with coastal forests are less likely to suffer death and destruction from tidal flooding and storm surges because they mitigate the hazard impact. As a result of this experience, and after an ecological assessment, environmentally responsible governments opt for forest/vegetation preservation and controls on any proposed development that would affect them [22]. Data from flume experiments indicated that submerged breakwaters seaward of a seawall would significantly reduce the impact intensity of incoming waves on a seawall [23].

3.3.2   Establishing Seawall Height

The height of seawalls varies with location based on still water level, wind speed and dominant direction, wave height, and a 50-year return period (from 50-year average) [24]. It should also take into account the history of heights of recurrent storm surges where applicable and the likelihood of sea level rise projected for the future (e.g., by 2050 and later) [25, 26]. For example, a modeled analysis of the effectiveness of seawalls to mitigate tsunami impacts for four locations in eastern Japan suggests that $a > 5$ m (~17 ft) high wall would reduce death and damage. Conversely, $a < 5$ m high seawall would increase the possibility of death and damage because it could encourage development in high risk vulnerable coastal locations [22]. With respect to modeling, there is the probability that physical model trials at a small scale can give qualitative data at best if a $c =$ scaling law is used and give bad data at worst [27].

That a well-placed high seawall can mitigate the impact of a tsunami was demonstrated at Pondicherry, SE India when a 2004 Indian Ocean tsunami that coincided with high tide struck 25 km (15.6 mi) off the coast. Pondicherry was protected by a 2 km (1.25 mi) long seawall 9 m (29.7 ft) above sea level, fronted by granite rubble. It was built in 1735 and maintained in good condition to the present. The tsunami at Pondicherry reaches a height of 8 m (26.4 ft) and the city escaped a major impact except for 25 people on a promenade who were killed [28]. As a result of the

2011 tsunami that followed a 9.0 Richter scale earthquake that hit northeastern Japan with the loss of more than 18,000 lives, destruction of homes, property and nuclear reactors that released dangerous levels of radioactivity that required the evacuation of areas within a 20 km radius of the reactors, the Japanese government reacted perhaps with a knowledge of Pondicherry. The government constructed the 395 km (245 mi) of seawall cited earlier that reached 12.5 m (41 ft) above sea level in some coastal locations [18]. In some areas, breakwaters constructed offshore of the Japanese seawall are expected to slow down a tsunami thus giving people more time to evacuate after an alert from the ocean-based tsunami warning system.

3.3.3   Determining Seawall Impact on Ecology

Seawalls affect ecological conditions by curtailing the natural landward migration of the ocean waters as sea level rises, and by changing the pattern of longshore drift that moves beach sand along a shoreline, thus reducing the width (area) of intertidal habitats. Longshore currents strip sand from a seawall front and deposit it where the seawall structure ends. These processes literally squeeze habitats out by trapping them between a rising sea level and an immovable hard defense with a resulting loss of biodiversity and organism abundance [17]. This can affect habitat value as an economic resource because natural resources a habitat can yield are lost (e.g., harvest clams) together with ecosystem functions (e.g., recycle wastes) that benefit people and their activities in this nearshore environment. It is essential that seawall placement be determined by a team of experts including bioscientists, marine geologists, and coastal engineers who will strive to preserve the intertidal biodiversity and complexity, and organism abundance as much as possible. This while maximizing the security of onshore (urban) populations, personal property, basic and critical infrastructure, and economic assets that provide employment and municipal tax income (e.g., beaches that attract tourism).

Researchers made a meta-analysis study that compared the biodiversity and organism abundance at natural shorelines with those at engineered defense structures. They found that at seawalls there was 23% less diversity and 45% fewer organisms vs. natural shoreline. Breaking this down further showed biodiversity at seawalls was lower by 66% for flora, by 20% for benthic infauna (life that burrows into soft sediment: polychaetes [worms], bivalves [clams]), by 52% for birds, and by 24% for nekton. Similarly, abundance at seawalls was lower by 66% for benthic infauna, by 71% for birds, and by 56% for nekton. This contrasts with results for riprap revetments (sea dikes) and for breakwater where there was no statistical difference in biodiversity or abundance in the organisms except for a 39% reduction in flora biodiversity at riprap revetments and a 39% greater diversity of nekton at breakwaters [29]. Non-burrowing benthic fauna such as crabs and fish were not assessed.

3.3.3.1 Chinese Ecological Experience and Response

China has employed a "get more land concept" to expand its economy by enclosing sections of coastal wetlands with seawalls. This land reclamation method has been used in the past by several nations but not to the extent used by China. From 1950 to 2000, 50% of the nation's mainland coast was (sea) walled off from the ocean. The reclamation continues to the present albeit at a slower pace. The reclaimed lands are the sites of aquaculture ponds, agricultural land, factories/industries, residential development to support urbanization, and salt ponds. This has resulted in an international decline in shared biodiversity and ecosystems services [30, 31].

Ecologically, natural intertidal wetlands include an upper and part of a middle intertidal zone with abundant growth of saltwater marsh plants that support migratory waterbirds when they stop to rest, feed in the flats to be ready for the next leg of a flight, and sometimes to breed. A part of the middle zone and the lower intertidal zone is unvegetated but has a high density of benthic organisms in the mud that supports biomass development there [31].

China's coastal area represents 13% of its land mass but produces 60% of the national GDP. Drivers of reclamation of intertidal zones at the expense of ecosystem degradation and loss of biodiversity were to move profitable development ahead to increase China's GDP growth. There was a general lack of management at the local and national levels of government with little attention being focused on what would happen to associated ecosystems. This, together with lack of environmental protection laws or the strong enforcement of existing laws, led to over-reclamation of intertidal and shallow ocean environments with their loss along with the loss of socio-economic and ecological benefits these ecosystems imparted to the nation. Loss of ecosystems services costs China an estimated US$31 billion or 6% of the income China gets from the ocean. The lost or diminished services include storm protection (loss of salt marsh wetlands and mangrove forests that intensifies the effects of extreme weather), water purification (remove pollutants from factories and antibiotics from aquaculture), and food production. The coastal region produces 28 million tons of fishery products that represent 20% of the world total from this area. The reduction of biomass from the intertidal zone and shallow sea, from changes in abundance of common species, and from ecosystems services could cost the Chinese economy US$177 billion annually from the loss of the Yellow Sea tidal flat habitat alone [30, 31].

The Chinese government has evaluated the concerns other nations have about their reclamation program and biodiversity conservation. A move towards an overseer agency instead of strictly local control should benefit coastal section development and environmental legislation enforcement. In a positive move, the Chinese government established 35 national nature reserves in the coastal region and 14 tidal flat systems have been designated as World Heritage Sites [30].

3.3.3.2   Seawalls as Microhabitats

Many seawalls are made of steel reinforced cast concrete. In order to encourage microhabitat marine growth and also present a wall that is designed to dissipate a wave energy loading, mitigate storm surges, and sap the force of tsunamis, the seaward face of a wall should be rough textured to encourage development of sea life habitats and microhabitats. Similarly, rubble that is put in place to form a toe at the sea floor protects it from erosion and undermining has a texture that serves the same purpose of encouraging habitat development.

3.3.4   Porous (Perforated)/Slotted Precast Concrete Seawalls

For more than four decades, marine coastal engineers have been studying the degree to which a porous/perforated (slotted) seawall and its width (cross section thickness in published studies) can reduce wave reflection and increase energy dissipation where dynamic pressure is greatest at the still water level (hydrodynamic parameters). This has the purpose of preventing run up of waves and overtopping of a seawall as well as run down that can erode a seawall at its base, undermine it so as to weaken the wall against continuous wave impact and end in it being pushed over. The research has focused on vertical simulations of seawalls. The research has used analytical mathematical data and computer modeling plus observations and measurements from flume studies at university and government laboratories.

The data show that a porous/perforated (slotted) pre-cast concrete seawall and incident wave height are controlling factors of the wave reflection characteristics [32]. Wave reflection is given as a coefficient calculated from the incoming wave height decided by the reflected wave height. Experiments show that this coefficient and run up at a seawall decreases with depth relative to seawall height, relative seawall width, wave steepness, wave velocity, and, of course, the degree of slotting and how slotted seawall openings are distributed horizontally and/or vertically. In addition, as noted earlier, submergered breakwaters reduced the energy of incoming waves [23]. Wave height at the landward seawall decreases exponentially with increasing seawall width but increases as slot (perforation) percentage increases [33]. Obviously, wave velocity decreases significantly as wave propagates through a slotted seawall and can approach zero velocity at a landward seawall face. The wave reflection run down at a seawall ocean side can have a large negative velocity that can erode sediment there and undermine a seawall unless protected by a toe of rubble that slows and dissipates the reflective wave energy [33, 34]. The research efforts strive to come up with the preferred design for given sets (a wide range) of hydrodynamic measurements (e.g., wave heights and wave return periods) [35]. To the present, despite all of the favorable data from laboratory simulations and computer modeling published over the years, perforated/slotted pre-cast concrete seawalls have not been used or ocean tested as coastal defenses.

3.3.5 A New Concrete for Seawalls?

Earlier in this text, it was noted that seawalls and other hard coastal defenses are expected to last for 50 years before obvious degradation becomes a factor especially in steel reinforced concrete structures. Degradation is caused by marine water with its sodium chloride and magnesium sulfate content that over time seeps into the concrete and contacts and corrodes the steel supports weakening the structure. The weakening is further enhanced by wave current activity, back swash erosion at a seawall base, changing tides, and weather (temperature changes, exposure to sun, and extreme events). Yet, seawalls built by the Romans 2000 years ago still stand. The "why" of this was answered in a 2017 report that found that this was the result of the mixture of volcanic ash and quicklime Romans used to make the concrete for seawalls. The volcanic ash came from one quarry in Italy and contained the mineral phillipsite. When seawater contacted a Roman seawall (no steel supports), a rare chemical reaction took place that formed the authigenic mineral aluminous tobermorite that grew out of the phillipsite as tiny crystal blades. These blades served as micro-armor that filled the concrete pores and prevented infiltration by seawater and degradation of concrete by inhibiting fractures from forming, thus strengthening the Roman seawalls we see preserved today [36]. Replication of the volcanic ash and quicklime mixture that Romans used can be used for contemporary seawalls. This would be a great advance and likely reduce seawall construction, monitoring, maintenance, and repair costs.

3.4 Surge Barriers

An innovation by Dutch marine coastal engineers to prevent flooding of Rotterdam and protect the port, the busiest in Europe, from North Sea storm surges was the construction of a steel storm surge barrier to block surge waters from the waterway that accesses the port and city. The barrier consists of two moveable arms with hollows suspended from them. They are dry docked on either side of the waterway and can extend from land and lock together to form the barrier. When dry docked they leave a 360 m (1188 ft) wide shipping channel. Each arm is 237 m (780 ft) in length. When the arms are extended, the hollows fill with water and settle to the channel floor forming the barrier. The barrier presents 22 m (~72 ft) high surge doors that are 15 m (~49 ft) thick. It was constructed from 1991 to 1997. The cost was given as €750 million. Its closure is fully automated when North Sea water level is 3 m (~10 ft) above normal and takes about 90 min. An 8 h advisory of the closure is sent to ships at sea waiting their turns to enter the port. The surge barrier has been closed only once in 2007 in response to a tidal surge and is tested once a year. The system was improved in 2005 and is improved as technology dictates [37]. There is a 25.4 km (15.8 mi) flood protection barrier for St. Petersburg, Russia, from Baltic Sea storm surges. The barrier includes a 200 m (660 ft) wide main navigation channel (S1) that has two 110 m (363 ft) arms that automatically move out from each shore dry dock chamber and link up and block the channel when sea level rises to more than 1.61 m (5.3 ft) above normal sea level. The operation parallels the method used in the Rotterdam Surge Barrier. The gates were closed in 2011 when the Baltic Sea level rose by 2.81 m (9.2 ft) [38].

A flood barrier in the Thames River was completed in 1982 to protect an area of 125 km², (~48 mi²) including greater London from tidal flooding and surges driven by North Sea storms up the river. Unlike the surge barriers described above, the Thames barrier sits on the river bed and is hydraulically raised as needed. The barrier spans 520 m (1716 ft) and comprises ten steel gates spaced so as to allow free flow of river traffic. The gates are 20 m (66 ft) in height. Raising all the barrier gates takes one-half hour. The system is not automated but rather started by a controller depending on computer signals giving the river heights at spring tides at the Thames estuary entrance, on the height of a tidal surge, or on the tidal height as it passes over a down river dam (weir) [39]. The cost when built was £534 million (US$776 million), or £1.68 billion (US$2.2 billion) in 2018 currency. It is tested monthly. The barrier has been closed to prevent flooding 182 times from 1982 to February, 2018.

References

1. Linham MM, Nicholls RJ (2011) Sea dikes. ClimateTechWiki. www.climatetechwiki.org/content/sea-dikes (scroll down for text)
2. Lenk S, Rynski D, Heidrich O, Dawson RJ, Kropp JP (2017) Costs of sea dikes—regressions and uncertainty estimates. Nat Hazards Earth Syst Sci 17:765–770
3. Mughal A (2018) Personal communication from Chief Technology Officer
4. Breteler MK (2016) Study of a new type dike revetment. www.deltares.nl/en/news/study-of-new-type-revetment
5. Hillblock the new standard in coastal protection. https://www.hillblock.com/site/en/nieuws/detail/20070051.html
6. Hill D. Narrated video. http://www.startupholland.tv/watch/147
7. CIRIA, CUR, CETMEF (2007) The rock manual. The use of rocks in hydraulic engineering, 2nd edn, C683, CIRIA, London, 1255 p. Contains design of marine structures, pp 773–908. Rubble mound breakwaters, pp 778–823. Rock protection to port structures, pp 823–835
8. Li H, Sanchez A, Wu W, Reed C (2013) Implementation of structures in the CMS (Coastal Modeling System). Tech Note ERDC/CHL CHETN-IV-93. US Army Corps of Engineers, Vicksburg, MS, 9 p
9. de Graauw A (2014) A long term failure of rubble mound breakwaters. J Mediterr Geogr. http://journals.openedition.org/mediterranean/7078
10. Jackson JL, Harley MD, Armaroli C, Nordstrom KF (2015) Beach morphologies induced by breakwaters with different orientations. Geomorphology 239:48–57
11. Montanari A (2017) Lecture: design of breakwaters. Online Search: Montanari, Design of breakwaters. www.albertomontanari.it
12. Sciortino JA (2010) Fishing, harbour planning, construction and management. FAO Fisheries and Aquaculture Technical Paper No. 539, Rome, 337 p. https://fao.org/decrep/013/i883e/i883e09.pdf, Chapter 7. Breakwaters, pp 87–196. Chapter 9. Construction materials, pp 133–179 (especially, pp 175–177, rocks)

13. Adams S. The pros and cons of breakwaters. www.hunker.com/134255855/pros-cons-of-breakwaters
14. Ruoi P (2017) Floating breakwaters. www.coastalwiki.org/wiki/Floating_breakwaters
15. Takagi H, Kashihara H, Esteban M, Shibayama T (2011) Assessment of future stability of breakwaters under climate change. Coastal Eng J 53:21–39. https://doi.org/10.1142/50578563411002264
16. Lorenzoni C, Postacchini M, Brocchini M, Mancinelli A (2016) Experimental study of the short-term efficiency of different breakwater configurations on beach protection. J Ocean Eng Marine Energy 2:195–210
17. Linham MM, Nicholls RJ (2010) Technologies for climate change adaptation: coastal erosion and flooding. TNA Guidebook Series. UNEP/GEF, Roskilde, Denmark, 150 p. http://techaction.org/
18. Jacobs S (2018) Business Insider. A series of comments on Japan's investment on constructing extensive seawalls after the 2011 Fukushima tsunami caused disaster. http://www.businessinsider.com/japan-seawalls-cost-12
19. Garfield L (2018) Manhattan plans to build a massive $1 billion wall and park to guard against the next inevitable superstorm. Business Insider. http://www.businessinsider.com.au/manhattan-plans-to-build
20. Bender E (2017) With storms intensifying and oceans on the rise, Boston weighs strategies for staying dry. Spirit of Change Magazine, Nov. 27
21. Department of Environment and Climate Change NSW and Sydney Metropolitan Catchment Management Authority (2009) Environmentally friendly seawalls: a guide to improving the environmental value of seawalls and seawall-lined foreshores in estuaries. Sydney, Australia, 27 p
22. Nateghi R, Bricker JD, Gulkema SD, Bessho A (2016) Statistical analysis of the effectiveness of seawalls and coastal forests in mitigating tsunami impacts in Iwate and Miyagi prefectures. PLoS One 11(8):e0158375. https://doi.org/10.1371/journal.pone.0158375
23. Abozaid AAEM (2014) Using porous seawalls to protect the coasts against sea level rise due to climate changes. M.S. Thesis, ZagaZig University, Egypt, 130 p
24. Yang Y, Hu X, Li Z (2015) The conditional risk probability-based seawall height design method. Int J Naval Architect Ocean Eng 7:1007–1019. https://doi.org/10.1515/ijnaoe-2015-0070
25. Intergovernmental Panel on Climate Change (IPCC) (2014) Climate change 2014. Synthesis report. IPCC, Geneva, 151 p
26. DeConto RM, Pollard D (2016) Contribution of Antarctica to past and future sea level rise. Nature 531:591–597
27. Kraus NU, McDougal WU (1996) Effects of seawalls on the beach. Part A. An updated literature review. J Coastal Res 123:691–701
28. Sheth A, Sanyal S, Jaisual A, Gandhi P (2006) Effects of the December 2004 Indian Ocean tsunami on the Indian mainland. Earthquake Spectra 22:S435–S473. Earthquake Engineering Research Institute
29. Gittman RK, Scyphers SB, Smith CS, Neylan IP, Grabowski JH (2016) Ecological consequences of shoreline hardening: a meta-analysis. Bioscience 66:763–773. https://doi.org/10.1093/biosci/biw091
30. Ma Z, Melville DS, Liu J, Chen Y, Yang H, Ren W, Zhang Z, Piersma T, Li B (2014) Rethinking China's new great wall. Science 346:912–914. https://doi.org/10.1126/science.1257268
31. Choi C-Y, Jackson MV, Gallo-Cajiao E, Murray NJ, Clemens RS, Gan X, Fuller RA (2017) Biodiversity and China's new great wall. Diversity and distributions. Wiley Online Library. Unpaginated https://doi.org/10.1111/ddi.12675
32. Zhu S, Chwang AT (2001) Investigations on the reflection behavior of a slotted seawall. Coastal Eng 43:93–104
33. Koraim AS, Heikal EM, Zaid AM (2014) Hydrodynamic characteristics of porous seawall protected by submerged breakwater. Appl Ocean Res 46:1–14

34. Karim MF, Tanimoto K, Hieu PD (2009) Modelling and simulation of wave transformation in porous structures using VOF based two-phase flow model. Appl Math Model 33:343–360
35. Nassar K, Negm A (2016) Predictive formulae for estimating of wave hydrodynamic parameters in front of sea walls. In: 19th international water technology conference, IWTC19, Sharm ElSheikh
36. Jackson MD, Mulcahy SR, Chen H, Li Y, Li Q, Cappelletti P, Wenk HR (2017) Phillipsite and Al-tobermorite mineral cements produced through low-temperature water-rock reactions in Roman marine concrete. Am Mineral Suppl 102:1435–1450
37. ABB Communications (2012) Protecting the Netherlands from storm surges and flooding. Discussion of the Maeslant Barrier Rotterdam. www.abb.com/seitp202/238f468a52449bb3c1 2579aa00452
38. Hunter P (2012) The St. Petersburg flood protection barrier: design and construction. Presented at the CET MEF PIANC, Paris, 9 p. https://eprints.hrwallingford.co.uk/603/1/ HRPP569_The_St_Petersburg_Flood_Protection_Barrier
39. Environmental Agency, UK (2014) Updated 2018. The Thames Barrier. www.gov.uk/guidance/the-thames-barrier (includes video)

Chapter 4
Coastal City Flooding

4.1 Introduction

Flooding is a natural hazard that can be a not-too-threatening event or a deadly one. In more violent events, floodwaters may move in great volumes, driven by gravity, and rage through highly and densely populated regions or rural areas killing and injuring people, destroying homes, roads, bridges and other infrastructure, and crops as the waters follow a pathway out of the flood zone. Flooding can also be an anthropogenic caused hazard.

4.1.1 Sources of Floods

Flooding in coastal cities often originates from tropical storms with wind velocities <74 mph (~119 km/h) and the more powerful hurricanes (typhoons, monsoons) with wind velocities 75 to 157+ mph (~252 km/h) that track from oceans inland. These storms can deliver heavy rains as they move across a metropolitan area. If a storm front stalls, it may rain down several inches on an urban center in a short time or be a sustained rainfall over a longer period of time until the storm continues inland where rainfall in drainage basins can be another source of flooding. Flooding in coastal cities may have a greater impact on edge neighborhoods with shanty-towns/slums that often occupy low-lying urban areas putting them at high risk from rising, rushing waters.

Coastal cities are susceptible to flooding from high and spring tides that may either passively rise and inundate low-lying areas of urban centers or be wind driven and more rapidly flood a city. This flooding can be defended with sea dikes as described in Chap. 3 but when defenses do not exist, or are not high enough, or are not maintained, the susceptibility to inundation increases and can result in serious economic problems. For example, Ho Chi Minh City, Vietnam, and Jakarta, Indonesia, are low-lying coastal mega-cities that regularly flood during high and spring tides. The flood condition from tides is worsening because both cities are suffering subsidence as the result of over extraction of freshwater from underlying aquifers and added to by sea level rise. This affects millions of their citizens today that with continued urbanization will grow to millions more in 2050. This costs their respective economies tens of millions of dollars annually to clean and repair damage to the flooded areas and will continue to do so until proper flood defenses are raised [1, 2].

Coastal cities can be flooded with storm surges driven inland by high velocity winds from tropical storms or hurricanes (typhoons, monsoons).

The storm surge hazard occurs mainly in Asia (South, South East, and East) and to a lesser degree in the southern area of the northern hemisphere Atlantic Ocean (in the Caribbean and Southeastern-South Central United States), in Eastern Mexico from the Gulf of Mexico, and in Europe from the North and Baltic Seas. Storm surge flooding can be a major threat to the well-being of citizens, property, and infrastructure. From 1900 to 2015, storm surges killed over 967,000 people for an annual average of more than 8300 deaths, with >98% in South, South East, and East Asia. Bangladesh, formerly East Pakistan, was impacted the most especially in 1970 by the Bhola Cyclone (~monsoon) driven from Bengal Bay by 60 mph winds and a reported 34 ft (10 m) surge that killed between 200,000 and 500,000 people. A 2018 paper reported that in recent years the mortality rate from such storms had been decreasing because of education, better warning communications from meteorological agencies, places to which people can safely evacuate, and emplacement of defenses [3].

However, during mid-March 2019, Cyclone Idai (~monsoon, typhoon) with heavy rains and winds reported at upwards of 175 km/h (110 mph [Category 2–3 storm]) tracked from the South Indian Ocean off the coast of Madagascar onto Mozambique's low-lying port city of Beira, (population ~534,000). The cyclone drove a storm surge of 14 ft (4.4 m) onto the city flooding it and together with the winds wreaked death, injury, damage, and destruction in its path. Much of low-lying areas of Beira were destroyed. A 7 mile (11.2 km) drainage canal system and water retention basin built after 2012 with US$120 million support from the World Bank to protect these areas was overcome by the storm and should be reworked to cope with future massive flooding. The cyclone continued on a path inland onto parts of Zimbabwe and Malawi impacting them with winds and floods from torrential rains that added to flooding from the earlier week of heavy rain across southeast Africa in these two countries before the Cyclone Idai made land. There have been more than 1000 deaths, thousands injured, homes destroyed, and hundreds of thousands displaced. Crops were lost so that post flood food security is an issue that will have to be resolved by donations from other nations. Because of poverty and lack of infrastructure or destruction of infrastructure that was in place, and loss of clean water supplies and sanitation systems, the World Health Organization is working to stem the outbreaks of disease supplied the affected populations with the cholera vaccine (more than 975,000 doses) and other medicines to help stem outbreaks of cholera, typhoid, and the endemic disease malaria (see Chap. 6, Sect. 6.1.1). Depending on the final total of lives lost, people injured, and devastation of homes, other structures, and infrastructure, this may be the deadliest and economically most costly storm to date that has hit southern Africa. Even considering that on the average, 1.5 cyclones track to the 2470 km (1535 mi) Mozambique coastline annually, rebuilding the port of Beira should include consideration of construction of sea defenses discussed in Chap. 3 that can mitigate the effects of such storms in the future.

Additionally, it would behoove a rebuilt Beira and other coastal cities and low-lying regions that are seasonally at risk of cyclone (monsoon, hurricane, typhoon) impact, and with less than suitable coastal defenses, to follow a program carried out by India after the 1999 Odisha cyclone killed 10,000 people, the majority from the storm surge and flooding. India developed improved storm forecasting and tracking models, instituted public awareness sessions, and practiced evacuation plans that included 45,000 volunteers that supported 9000 cyclone shelter, and 7000 kitchens to feed evacuees. Bangladesh did much the same with 4000 cyclone shelters stocked with safe water, dried foods, and medical supplies. When Cyclone Fani was tracked moving to India's poor, densely populated Odisha coastal region as a Category 4 cyclone at the end of April and beginning of May, 2019, 1.2 million people were evacuated to shelters in 24 h while 1.6 million people were evacuated to shelters in Bangladesh. When the cyclone hit on May 2, and tracked inland less than 50 deaths were reported by May 6.

As noted in previous paragraphs, storm surges may drive high waves onshore and inland. Surge height is influenced in part by the topography (hydrography) of the sea floor shoreward and configuration of the coast such as with bays/inlets and promontories and inland topography. Hurricane Sandy that impacted New York City in 2012 caused a surge ~14 ft (~4.2 m) and the surge at New Orleans from Hurricane Katrina in 2005 was higher than 25 ft (8 m). In 2018, a lesser storm surge of ~5 ft (~1.6 m) driven by 45 mph winds hit Venice, a port city essentially below sea level, and flooded 77% of the city. Chap. 3 discussed how seawalls or moveable barriers can mitigate or eliminate flooding from storm surges. A caveat here is that seawalls should have one way valve portals through which surge floodwaters can readily recede back to the ocean. Flooding from these storms, apart from surges, is more severe and longer lasting when surge waters are fortified by heavy or torrential rains directly onto coastal cities as storms track slowly inland or if a weather front stalls over them.

Coastal and inland cities worldwide can also suffer flooding when there are steady sustained long-term rains or heavy/torrential downpours in drainage basins (catchments) they are part of. Drainage basins are areas from which all rainfall or snow/ice melt flows to a single river (main trunk) or to a collection of rivers that follow the topography in their flow towards an ocean or estuary. The waters' flow path(s) may be through or near highly and densely populated urban centers sited at or close to the marine or estuarine shore. When the volume of water in a basin's streams or rivers is greater than the carrying capacity of their channels, over bank flooding ensues. The depth of floodwater, the width of a river's floodplain (or reach of the flood), and the velocity of its flow results in a flood force are dependent on two factors. These are the volume of water runoff that enter a channel (not absorbed by soils) and the local/regional topography. Hydrologically, a drainage system includes subsurface water that moves in aquifers gravitationally down dip (slope).

Flooding can be a repeat event that may occur regularly in some areas or in other statistical time frames that may be categorized as a once in a decade event, or other time periods such as floods that recur once in 100 years, or once in 500 years, or more. The area inundated by a flood is referred to, for example, as a 100-year floodplain or a 500-year floodplain. Zoning laws should prohibit habitation or industrial/manufacturing development on a 100-year floodplain since the statistics here are estimates so that floods may recur more often than indicated by the analytics.

4.2   Flood Warning, Flood Control

A citizenry in coastal or inland cities, alert to a warning and arrival time of an oncoming flood, can be given enough time to respond in several ways previous to a stand-down condition or to a mandated evacuation. Citizens may sandbag openings to a building, board up windows, move items they can save to upper levels of a home and gather important papers and valuables before the time comes for a mandated evacuation. If effective flood defenses are in place, it may not be necessary to evacuate.

4.2.1   Flood Alert

For a first-line of defense for citizen safety when a flood event is in the offing, it is essential to have flood alert systems and plans in place to keep a population informed. For coastal cities, this is the job of a meteorological agency that tracks an incoming storm and projects when and where it will make land and its wind speed, the possibility of a surge, the coastal and inland areas threatened, and the amount of rainfall expected. On this basis, a government will determine whether it is advisable (or mandatory) to evacuate.

As a storm moves inland, urban centers in a drainage basin, including coastal cities near or in a river's flow towards an ocean or estuary will be at risk of flooding. Unlike most natural hazards, flooding in a drainage basin can be monitored with stream gauges coupled with computer system models that can predict when a flood condition will rise at a location in the basin and give reliable estimates of how high floodwater will rise as it overflows river banks. With this alert data put out by broadcast media and by police and fire departments and civil defense personnel, and smartphones, citizens can gather their important documents and valuables as they prepare for evacuation to safe centers that can receive them and attend to their basic needs until floodwaters recede.

4.2.2   Flood Defenses/Controls

To reduce the dangers of major flooding that originates from rainfall or snow/ice melt in a drainage basin, a city (coastal or inland) should have flood defenses in place. These may be flood control dams with spillways where a filled dam can release otherwise potential floodwaters to a safe location. Defenses may be levees and river walls that contain waters in river channels thus preventing over bank flooding. Another control to prevent or minimize flooding uses large pumping stations such as in the United States southeastern coastal cities, New Orleans and Miami. The pumping stations are strategically located to capture floodwaters from high and spring tides (Miami) or during a tropical storm or hurricane (New Orleans) and move them out to the ocean reducing risk for city populations. The failure of at least one pumping station during Hurricane Katrina in 2005 contributed to the New Orleans flood disaster. If defenses are not in place where there is a history of recurrent flooding, or are in place in some parts of a municipality but not others because of unplanned urbanization, it is imperative for citizens safety to plan for and implement their installation where they are lacking.

An important flood control strategy is to move floodwater through an area as fast as possible. To this end, it is essential to keep outflow channels free of debris that otherwise slows water outflow. This is often the case in unauthorized shantytowns/slums in coastal cities in less developed and developing economies. It may also be necessary to dig canals to carry floodwaters out of cities into safe areas or oceans.

Four costly options exist that will move a greater volume of potential floodwaters more quickly through a city's existing streams or rivers. Volume flow capacity of channels can be increased by deepening and/or widening a channel bed, and/or by installing concrete walls at river banks. Straightening a channel will move potential floodwater more rapidly through an at-risk area. These four methods individually or in a degree of combination can mitigate the impact of urban flooding. Maintenance to periodically remove newly deposited sediments from these channels is essential.

4.3   The Tsunami Factor: Not a Flood But a Seawater Threat

The ultimate destructive mass of seawater that can wreak havoc in coastal cities and smaller population centers is a fast moving, high energy tsunami wave that can rise to 10s of feet (several m) as it gathers height in its run towards shore and crashes into and onto a coastal area. Tsunamis originate mainly in the Pacific Ocean "ring of fire" tectonic active zone. A massive tsunami will roll into and flood an impact area rapidly killing and injuring people, destroying buildings and infrastructure and dragging bodies and masses of debris out to sea as it just as rapidly recedes only to return one or more times but with ebbing force each time. Great numbers of city people may be displaced by a tsunami as destruction of buildings, infrastructure, and industrial installations can be monumental, vis a vis Fukushima, Japan. Here, a

2011 tsunami roared onshore killing thousands, leveling buildings and other property, and destroying part of a nuclear facility complex that resulted in the release of radiation that put a large swath of land with homes, schools, and businesses off limits for habitation, displacing thousands of inhabitants many of whom remain in temporary housing in 2019. Today, low-level radiation is still being released via groundwater into the Pacific Ocean offshore of northeast Japan.

As noted previously, tsunamis associate with the "ring of fire" bordering the Pacific Ocean tectonic plate that gives rise to earthquakes and volcano eruptions. Tsunamis occur after a sea floor earthquake when one side of a major fault rupture on the seafloor moves up or down several feet (or meters) relative to the side from which it breaks initiating a cascade of seawater heading to a coast. Some tsunamis that rise several feet (or meters) high as they move landward pose a grave danger to coastal populations, while those a few feet high may not. The tsunami caused by the 2004 9.1 Richter scale earthquake off of Sumatra, Indonesia, killed more than 200,000 people mainly on Sumatra but also in other nations' coastal populations. Tsunami warning systems were in place in some areas of the Pacific Ocean were not then in place in this area. They are now in place here and in other areas where tsunamis are most likely to occur.

Tsunami warning systems are comprised of deep-sea buoys fixed to the sea floor. They are electronically equipped to determine the epicenter of an earthquake, the height of seawater above the sea floor as it moves from one buoy to another, and direction and velocity of the water mass (up to 500 mph (800 km/h)) depending on the depth to the sea floor. The data the warning systems generate are computer processed rapidly (~5 min) and signals are sent to shore stations telling when and where a tsunami will hit land and what height it is estimated to reach. If the shore alert systems are functioning, they can give people warning time to evacuate to higher inland locations as necessary. A relatively close to shore earthquake can generate a tsunami for which a warning would not give citizens much time to evacuate inland to higher ground. Human error can negate a tsunami warning and alert system with disastrous results. Again, in the "ring of fire," a 7.5 magnitude ocean earthquake in September, 2018, shook the city of Palu on the Indonesian island of Sulawesi causing great physical damage and giving rise to a tsunami that killed 2500 people, injured many more, and displaced 18,000 of the city's 300,000 population. Here, the tsunami warning system did not function as planned. Blame was put on the government agency that lifted the warning about 30 min after the earthquake happened, a deadly mistake, or on texted warning messages that did not go out because cellphone towers were downed by the earthquake. This was a learning moment for those responsible to review and revise as necessary their tsunami warning capabilities especially for its redundancy in case of a primary system failure.

On rare occasions, tsunamis can be generated by volcanic activity. On December 22, 2018, the island volcano Anak Krakatau that had been in eruption phase since June erupted and likely caused undersea landslides or landslides from the volcano itself into the sea that triggered a tsunami. It impacted the coast and ran up the shore 20–30 m (65–100 ft) as the 1–2 m (3–7 ft) rapidly moving wave rose onto beaches filled with people on a festive Saturday night around Indonesia's western Java and southern Sumatra islands without warning 24 min after a strong eruption. The tsunami killed at least 429 people, with 154 missing and 1485 injured as of Christmas, 2018. It damaged or destroyed nine hotels, 600 homes, vendor stalls, and 400 boats or ships. There is a network of detection buoys keyed to earthquake detection that would likely respond to a major landslide event, but it had not been functioning since 2012 due to budget shortfalls and perhaps vandalism. One hopes that the Indonesian government will find funds to repair this system before another tsunami disaster impacts the country. Clearly, this is an important event because it emphasizes another cause of failure of a tsunami warning system that can affect coastal cities in the Pacific Ocean "ring of fire" and emphasizes that a tsunami does not have to reach several meters in height (as in Fukushima, Japan) to be a killer wave.

4.4   Flooding by Dam Failure

In some cases, floods happen because of dam failure. There are several reasons for building dams [4]. They are emplaced to serve as reservoirs against drought, to impound water for irrigation, to generate hydroelectric power for navigation, for diversion of a river, and as noted previously, for flood control defenses. In recent times, only one coastal city with a high population has been affected by flooding from dam failure. The southern area of the eastern Spain port city of Valencia (population 780,800) was flooded from the 1982 failure of the Tous Dam. Nonetheless, an awareness of this possibility is important to coastal cities, especially those with high populations that continue to increase. As noted earlier in the text, the 2018 global urban population is 4.1 billion people of the worldwide population of 7.6 billion. These figures are projected to increase in 2050 to 6.9 billion people in cities with 3 billion in rural locations [5]. The coastal cities population is projected to grow from about 1 to 1.7 billion during that time. This makes added water security essential for many cities because of more people to supply and projected global warming/climate change that puts some populations at increased risk of droughts [6, 7]. A good portion of the growth will be in or near coastal cities. These growing cities may plan to build a reservoir dam in the future to assure a reserve water supply against drought, or a dam for flood control. Siting dams in topographic regimes where waterways do not flow close to such cities is essential.

4.5   Addendum: Rain Stimulated Landslides During/Post Floods

Finally, there are incidents yearly where many people die because the same sustained, or heavy, or torrential rainfall that can cause flooding can also trigger landslides. This is the result of weakening slope resistance to the pull of gravity on a hill or mountain underlain by sedimentary earth materials, especially where strata dip to a low area or valley face. Resistance to the pull of gravity is compromised by the added weight of water that seeps into a hill or mountain and the lubrication of earth materials. When resistance is overcome, a sedimentary mass detaches and slides in downslope crushing and burying people and structures in the slide path. Planning in coastal cities that have hilly terrain and seasons of heavy rainfall has to consider how to protect people and infrastructure from landslides since there is no electronic warning system to alert people that they are in the path of an incipient landslide. Potential for a landslide from a hillside of soil covered, weakly consolidated sedimentary rocks can be mitigated by installing a drainage system with pipes that collect and direct seepage to safe discharge points. This reduces the pull of gravity on the rocks and spread of lubrication that otherwise could destabilize the hillside. Essentially eliminating the possibility of a landslide may be by sealing a hill that threatens important coastal urban and inland infrastructure so that rainwater or snow melt cannot seep in to destabilize it. Runoff here is directed to safe discharge locations. This is the case in Japan to protect the train route from Tokyo to visit Lake Ashi and Mt. Fuji (Fujisan).

References

1. Tran Ngoc TD, Perset M, Strady E, Phan TSH, Vachaud G, Quertamp F Gratiot M (2016) Ho Chi Minh City growing with water-related challenges. In: Water, Mega-cities and Global Change, 27p. eaumega.org/wp-content/uploads/2016/05/HCMC-MonographyEN.df
2. Abidin HZ, Andreas H, Gumilar J, Brinkmann JJ (2015) Study on the risks and impacts of land subsidence in Jakarta. In: Proceedings of International Association of Hydrological Science, vol 372, pp 115–120. https://doi.org/10.5104/pihas-372-115-2015
3. Bouwer L, Jonkman SN (2018) Global mortality from storm surges is decreasing. Environ Res Lett 13:s014008. https://doi.org/10.1088/17489326/aa98a3
4. Civil Engineers Forum (2016) 8 Different types of dams. civilengineersforum.com/common-types-of-dams
5. World Population Data Sheet (2018) Population Reference Bureau, Washington, DC
6. Intergovernmental Panel on Climate Change (IPCC) (2014) Climate change 2014. Synthesis report. IPCC, Geneva, 151p
7. Intergovernmental Panel on Climate Change (IPCC) (2018) Global warming of 1.5°C. Special report. IPCC, Geneva

Chapter 5
Physical Care: Lessening Impacts from Other Natural Hazards

5.1 Minimizing Earthquake Impacts on At Risk Coastal Cities

Earthquakes happen by the thousands each year but only a small number annually release the seismic energy that can kill and injure people and damage and destroy structures and infrastructure. A regional recurrence interval for major earthquakes as well as the probable Richter scale magnitude can only be very roughly estimated (guesstimated?). For example, the United States Geological Survey gives probabilities of high magnitude earthquakes hitting the Los Angeles and San Francisco areas in 30 years. For Los Angeles, there is a 60% probability that there will be an earthquake of 6.7 (destructive) magnitude. For San Francisco, there is a 72% probability that there will be a 6.7 magnitude earthquake. Whatever the weakness in prediction and the inability to prevent an earthquake, locations must prepare so as to limit an earthquake impact on citizens and assets. The probabilities that greater magnitude earthquakes will hit the two California areas are given in Table 5.1. Other countries at risk of earthquakes in coastal cities make similar estimates. For example, 2012 data from Tokyo University Earthquake Research Institute gave a probability of 70% for a magnitude 7 earthquake impacting Tokyo in 4 years. It did not happen. A prediction of a 98% probability in 30 years, 2042, has 23 years to occur or not.

5.1.1 Examples from Circum-Pacific Tectonic Belt

Many coastal cities with high and ofttimes dense populations are at risk from the shaking, jarring, and rolling motions of earthquakes that can kill hundreds to thousands of citizens as buildings collapse and infrastructure destroyed. The

F. R. Siegel, *Adaptations of Coastal Cities to Global Warming, Sea Level Rise, Climate Change and Endemic Hazards*, SpringerBriefs in Environmental Science, https://doi.org/10.1007/978-3-030-22669-5_5

Table 5.1 Calculated probabilities that major earthquakes of given Richter magnitudes will hit the Los Angeles and San Francisco areas in 30 years [1]

For Los Angeles area
60% for an earthquake measuring 6.7
46% for an earthquake measuring 7.0
31% for an earthquake measuring 7.5
For San Francisco area
72% for an earthquake measuring 6.7
51% for an earthquake measuring 7.0
20% for an earthquake measuring 7.5

tremors originate from faults related to tectonic plate movement that suddenly rupture at kilometers (miles) beneath the ocean floor or faults that run through a city or close to a city, releasing seismic energy that may not be threatening or that can be powerful as a damaging, destructive force. As noted in the previous chapter, many at-risk coastal urban centers are located following the Pacific Ocean's "ring of fire" caused by plate tectonic activity that can release energy via earthquakes or volcanic eruptions. Some coastal cities threatened by earthquakes in this region are Tokyo, Osaka-Kobe, Jakarta, Manila, Christchurch, Santiago (Chile), Lima, Southern Mexico, Los Angeles, San Francisco, and Seattle. In the Middle East, cities such as Beirut and Istanbul (Constantinople) are populated coastal areas at risk from earthquake activity. In Asia, Dhaka is a coastal city that has experienced damaging earthquakes. In Africa, Alexandria has been impacted by earthquakes in the past. All efforts at mitigation of dangers from earthquakes should work towards minimizing such impacts on people and cities that may occur in the near or distant future.

Because there is no way to reliably predict the onset of an earthquake and no way to prevent an earthquake, the cities mentioned above, as well as those with an earthquake history at inland global areas, it is an absolute necessity to be prepared. This means adhering to protocols for the erection of new structures according to local building code, inspecting the work as building goes on to assure compliance, and retrofitting existing buildings to give them greater resistance to earthquake motions. It means being prepared to help citizens during (if possible) and after an earthquake with clean water, food, tent protection, and toilets as needed. Reestablishment of lost electricity is basic to recovery efforts. To this, readiness adds medical assistance on site as well as at hospitals and clinics, or field hospitals if needed, staffed and well supplied with medicines or with readily accessible stocks of them. Waste collection and its proper disposal is essential, especially of organic wastes in order to prevent an infectious disease (e.g., cholera) and its spread. Search and rescue teams and heavy equipment to remove rubble are the backbone of post-earthquake efforts.

5.1.2 Building Codes: Retrofitting-Post Earthquake Access

With respect to building codes, nations or municipalities in earthquake danger zones should develop or update their codes according to the norms set by the Japanese government or by the state of California, USA [2, 3]. These add expense to construction methods and materials but are investments that are warranted for the protection of people and property when an earthquake occurs. Retrofitting existing at-risk structures as determined from inspections by structural and civil engineers is costly but is an investment wisely made. Japan has retrofitted existing high rise buildings in Tokyo with base isolation equipment and other earthquake resistance technologies. At the same time, the Japanese government, confident in the base isolation technology, prepares these retrofitted structures to receive hundreds to thousands of earthquake victims with stored food, water, cots and blankets, toilets, and basic medications for 3 days. Jakarta, on the island of Java, is a mega-city with more than ten million inhabitants and metropolitan Jakarta has a population of ~28 million people, including large populations living in shantytowns/slums. Jakarta has more than 950 high rise buildings including 211 skyscrapers, 90 with heights >150 m (>500 ft) and 51 with heights >200 m (670 ft). The average age of these buildings is ~9 years. Fifty-two skyscrapers are in construction with some planned to reach a height of 300 m (>1000 ft). After major earthquakes on Java in 2017 and 2018 that shook buildings in Jakarta, the municipality undertook the inspection of more than 500 high rise buildings to determine what retrofitting against earthquakes, if necessary, should be done on a priority sequence.

Other methods to prepare citizens to counter the impacts of earthquakes are used [4]. One problem that can exist after an earthquake is lack of access to supplies to some populations because of a bridge collapse. Japanese engineers have developed a folding (like an accordion) light weight portable (by truck) bridge. They have built a prototype that can span 56 ft (~16.5 m) and can carry 5 tons (a small loaded vehicle). The bridge is equipped with hydraulics to extend the bridge and deck plates and can be deployed by three technicians in 10 min. The bridge weighs about 13 tons and is 9.5 ft (~3 m) wide [5].

5.2 Mitigating Wind Damage from High Energy Storm Events

The winds that drive storm surges that inundate inland regions wreak their own disasters on coastal/near coastal communities. Five categories of sustained wind velocities and the disasters each can cause have been defined by the Saffir-Simpson Hurricane (Typhoon, Monsoon) Wind Scale. Categories 1 and 2 have sustained winds of 74–95 mph (119–153 km/h) and 96–110 mph (154–177 km/h), respectively. They leave some damage to well-constructed frame homes and can snap branches off and/or topple shallow rooted trees downing power lines and blocking roads. Devastating to catastrophic damage is wrought by Categories 3 (111–129 mph [175–208 km/h]), 4 (130–156 mph [209–251 km/h]), and 5 (>157 mph [252 km/h]) with major to severe damage to framed housing by loss of roofing to wall collapse and their total destruction. Trees will be uprooted, power lines downed with extended electricity outages, and loss of water treatment capability can render residential areas uninhabitable for weeks or longer until repairs are made. Fortunately, the most destructive of these, Categories 4 and 5 are relatively rare.

What adaptations then can be undertaken to try to minimize the coastal city damage/destruction from the lesser and the more powerful winds as hurricanes (typhoons, monsoons) hit coastal cities? Roofs (e.g., shingles, tiles, slate, or even a roof deck) can be lifted up by the "vacuum force" of winds. A strong enough force (e.g., as just noted Categories 3, 4, 5) can rip off roofs. There is strength that can be added to a roof either during home construction or to an existing home that may minimize damage a roof might sustain from a high energy tropical storm. For example, during construction, nails, not staples, should be used to secure a roof deck made of high quality plywood to a home's rafters. Also, "hurricane" clips (straps) of galvanized metal should be used to connect the roof to a home's walls from the inside to better stabilize it. These may be difficult to use to retrofit an existing home because of the tight space where they are needed. The shape of a roof affects its stability under high wind velocity stress. A pyramid shaped roof is more stable than a gabled roof (inverted V shape) that is susceptible to being pulled off its deck. The stability gabled roofs can be improved in existing homes by installing wooden braces (2 × 4 in.) that attach the ends of gabled roofs to rafters. Vents at the ends of gabled roofs should be closed when a hurricane event is forecast in order to prevent wind pressure buildup from within that would abet roof detachment [6, 7].

Powerful winds can develop pressures that blow out windows and doors and allow invasion by flying debris driven by high velocity winds, debris that can hurt inhabitants who "ride out" storms or damage home interiors. This hazard is mitigated by boarding up the windows and doors with plywood. It is possible, as well, to replace regular window glass with hurricane impact glass that is 3/8 in. thick and glazed with the same film as used in automobile windshields that can break but not shatter. For existing housing or housing in construction, the installation of metal storm shutters to prevent the direct wind loss and losses from storm waters that otherwise could enter a home (or business) is a good investment because hurricane insurance premiums can be greatly reduced. In Florida, USA for example, hurricane insurance on a $150,000 house is $3000–$5000 annually but with metal storm shutters the premium is $1000–$3000 [6, 7].

5.3   Inland Anthropogenic Activities that Exacerbate Threats from Oceans, Extreme Weather, and Terrestrial Hazards

5.3.1   Discharging More Water from Aquifers Than Is Being Recharged

Chapter 2 emphasized that one of the results of extracting more water from an unconfined coastal aquifer than is being recharged is to reduce hydrostatic pressure and allow salt water intrusion into the aquifer. A second serious result that is having a damaging effect on many coastal cities is subsidence. Subsidence, the slow lowering of a portion of the earth's land surface, presents a grave problem in some high population coastal cities, especially in Southeast Asia such as Bangkok, Ho Chi Minh City, and Jakarta. Besides being a result of extraction of more water from aquifers than is being recharged, subsidence can be caused as well by the extraction of petroleum and natural gas from subsurface sedimentary strata. In both cases, the buoyancy strength added to sedimentary rocks by the fluids lessens if fluid is not replaced with the result that the subsurface strata slowly compact and the land surface lowers. Depending on how much fluid extraction takes place at different extraction wells in an aquifer or oil field, there may be differential subsidence from one area to another. Injection of water into aquifers or oil field reservoirs can slow or stop a subsidence and may allow minor rebound. As discussed next, loading at the surface can also contribute to subsidence. In many Chinese cities, including the important coastal cities of Shanghai and Tianjin, engineers attribute 70% of the subsidence to over withdrawal of aquifer waters and 30% to the weight of high rise buildings and skyscrapers.

5.3.2   Loading Weight at the Surface

As just noted, loading the surface with high rise and skyscraper buildings with the weight of buildings themselves plus the weights when furnished and occupied contributes to subsidence. For example, in Shanghai, the subsidence rate was 3.84 mm/year from 1980 to 1989. The 1989 population was 8.7 million people. As China's economic development progressed, the city's population grew and high rise buildings and skyscrapers were built to house citizens and companies and the rate of subsidence increased. From 1989 to 1995, subsidence was 9.97 mm/year and the 1995 population reached 11.1 million people. From 1995 to 2005, Shanghai's population grew to 17.1 million inhabitants and subsidence was 12.01 mm/year. Cumulatively, in the 25 year period from 1980 to 2005, Shanghai subsided ~21.2 cm (~8.4 in.) [8]. A seawall protects urban Shanghai from seawater flooding. To adapt to the problem, the government put controls on water withdrawal from the aquifer. Water demand was met by rerouting waterways to supply water to treatment plants that delivered safe water to buildings and by recharging the aquifer system with injection wells. This was designed to limit or stop subsidence from aquifer discharge. To prevent or limit overtopping of the Shanghai seawall from extreme weather storm surges abetted by sea level rise, the Chinese government was prepared and adapted by heightening the seawall.

Differential subsidence can happen as well when the bearing strength of a soil/sediment at a high rise (perhaps skyscraper) building site has not been well assessed either as to subsidence that would take place, the amount of subsidence that would be expected, or whether the compaction of earth materials underlying a foundation will be equal or differential. Nor has the benefit of setting a foundation through soil/sediment into bedrock that would give stability to a structure that has been thoroughly assessed and tested.

Subsidence can happen as well when the foundation of a planned high rise (perhaps skyscraper) building is not set in bedrock that would give the structure stability but instead is set in soil/sediment that may not have the bearing strength to sustain the weight of the structure and would likely compact. Civil engineers can assess whether this will take place at a building site, the amount of subsidence that could be expected, and whether any subsidence would be equal or differential so as to cause a structure to tilt. If subsidence would be the result predicted by material analyses, engineers would recommend setting the foundation in bedrock or possibly using a floating foundation. This latter is where a weight of earth materials underlying a building site is excavated to match the weight of a structure plus that of people and furnishings that would occupy it, thus maintaining equilibrium.

A recent example of a skyscraper building foundation not being set in bedrock and a resulting differential subsidence and tilting of the structure is the Millennium Tower in San Francisco. The Millennium Tower is a 58 story skyscraper condominium (645 ft [~196 m]) high. Construction began in 2005 and cost $350 million. Occupancy began in 2008. By 2018, the building subsided 17 in. (42.5 cm) and tilted 14 in. (35 cm) to the northwest. Instead of putting the foundation into bedrock at 200 ft (~61 m) depth, the foundation was set into packed sand and debris at 80 ft (~24 m) depth. The structure will continue to sink at 1.6 in. (4 cm) annually (satellite analysis) close to double the rate estimated by consulting engineers. Of very great concern is that during an earthquake in the San Francisco region, the soil could move out from under the heavy structure by liquefaction causing instability [9]. In 2018, engineers proposed a "fix" by drilling steel and concrete "micro piles" into bedrock to stabilize the tower and bring it back to upright [10]. An estimated cost for this solution would be between $200 and $500 million. Whether the investor group gives the "fix" a go ahead is not yet known. A question exists of why city building inspectors having information about subsurface conditions and the massiveness of the proposed tower did not require the foundation to be set in bedrock. More expensive? Yes. But at what later cost if investors go for a lower cost?

5.3.2.1 Bangkok

Bangkok is a coastal city with almost 16 million people. The city generates about 30% of Thailand's GDP. The mean elevation above sea level is 1.5 m (~5 ft) with the elevation decreasing to 1 m (~3 ft) moving south towards the coast. As in most cases of subsidence, the cause has been mainly from extraction of more water from multiple subsurface aquifers than was recharged and the subsequent loss of buoyancy pressure that gave strength to aquifer sedimentary rocks. Also, Bangkok has 700 structures with 20 or more floors and 4000 5-story structures. The added loading of weight at the surface from these buildings contributes to the subsidence [11].

From 1978 to 1981, subsidence in Bangkok averaged about 10 cm (4 in.) annually. From 1978 to 1999, the cumulative subsidence in some areas of the city reached 100 cm (1 m [>3 ft]). This intensified the risk of flooding that is the major natural hazard affecting Bangkok. For example, in 2011 Thailand was inundated by one of the worst floods in its history. Millions of people were affected and 800 killed. Bangkok was under water for months with the loss to the Thailand economy of $40.9 billion.

The repeated flooding, deaths, injury, and loss of property and economic input spurred the Thailand government to pass legislation that dealt with the problem by requiring a water management program that would slow down and ultimately arrest the subsidence. The Bangkok Water Management Commission put in force a mitigation program that would control extraction of groundwater from the multiple aquifers underlying the metropolitan area. The program reduces the permitted usage of registered wells. It requires that unregistered wells that extracted much of the groundwater that caused subsidence, register or be shut down. Importantly, the program promotes public awareness in groundwater conservation and does not permit groundwater development in the public water supply service area. In addition, the Commission established a groundwater tariff and a groundwater conservation tax [12]. In this way, water extraction from aquifers was controlled with an equilibrium slowly being established between discharge and recharge and a reduction of subsidence as measured from 2005 to 2010 to <1.5–2 cm (0.6–0.8 in.) annually [13]. A control on construction of multi-story and high rise buildings and infrastructure reduced the surface weight burden caused by earlier construction.

Aquifer recharge is also abetted by injection wells that draw water from the Chao Phraya River basin. If the subsidence continues even at 1 cm annually until 2050, the subsidence would be 31 cm (12.4 in.). At the same time, if sea level rises 3.3 mm annually this would be 4 in. (6.5 cm). Together, these changes would give a 16.4 in. (41.6 cm) relative change in coastal conditions that would intensify the flooding hazard from the sea (high and spring tides, storm surges) and from the river basin when torrential rains strike inland. To keep Bangkok functioning will require stopping subsidence, improving flood defenses for this mega-city, as discussed in Chap. 3, including pumping stations that move incoming flood waters out to sea as used in New Orleans and Miami, USA.

An encouraging study using persistent scatter interferometry with imagery from November 2007 to December 2010 detected uplift of the surface around Bangkok of 0.5–3 cm (1/4 to 1-1/4 in.). This portends well for Bangkok. A continuing rebound will likely be the result of the program that balances the volume of water allowed to be extracted against the recharge of unconfined aquifers underlying the city complemented by injection wells that added to the recharge volume and hence increases in buoyancy pressure [14].

5.3.2.2 Ho Chi Minh City

Ho Chi Minh City had ~10.7 million citizens in 2018 and continues to grow with a projected population of ~14 million people in a decade. This coastal city is 1–5 m (3+ to 16 ft) above sea level with a 3 m (~10 ft) average. The city accounts for almost 30% of the national industrial output. The most important hazard affecting the municipality is flooding during August through December from high tides, spring tides, and steady, long lasting or torrential rains. The flooding overwhelms low areas as well as important sections of the city center and abutting neighborhoods. Of 116 roads frequently flooded during high tides/spring tides, 79 are flooded because of subsidence. Ho Chi Minh City is subsiding at a rate of 1–2 cm annually with much differential subsidence within the city ranging from 10 to 15 mm annually at one area and from 5 to 20 cm annually at other locations. The subsidence is greatest where urbanization is highest. In all, the government has recently identified 56 areas susceptible to flooding for which they plan to develop flood defenses and control [15].

Subsidence in Ho Chi Minh is caused in great part from over use of aquifer water because of the increased demand from growing urbanization. Water is being pumped from recorded wells at a rate of 1 million m^3/day, a volume greater than the aquifer water recharge rate of 200,000 m^3/day. In addition, there are an unknown number of unrecorded wells. Safe water (mainly from the underlying aquifer) reaches 99% of the population and sanitation serves 94% of the population. As noted in a previous section, when the water discharge is not equal to the recharge, there is a reduction of buoyancy pressure that otherwise gives strength to the aquifer sediment that resists subsidence. An added cause of subsidence in Ho Chi Minh City is building and infrastructure construction that loads weight onto the surface that abets compaction of underlying soils and sediment. Construction has also covered 50% of what were aquifer recharge areas with concrete [16]. Building of high rise structures and skyscrapers has been increasing since about 2011. There are 43 skyscrapers built, 16 under construction, 15 approved, and 15 planned [17]. This parallels the subsidence causes in Jakarta and Bangkok.

Under existing conditions, Ho Chi Minh City suffers high tide flooding and citizenry is at high risk from storm surges. Continued subsidence abets encroachment from rising sea level, increased susceptibility to flooding by high and spring tides, from storm surges, and from water flowing downstream in the drainage basin after a rainy season tropical storm. The future bodes badly for Ho Chi Minh City because of the likely submergence of large coastal areas in two generations (e.g., 50 years). At a subsidence rate of 1 cm annually during that period and an annual sea level rise of 3.3 mm (likely to increase) would mean a water level 58 cm (~23 in.) above that in 2018. A 2 cm subsidence rate would bring the level to 1.16 m (46 in.) above that in 2018.

Clearly, this calls for several modes of action to minimize the problems that exist and that would intensify in the future. One line of action is to adopt a strict water management program as was done in Bangkok. This means limiting known well withdrawals and closing down unregistered wells with the aim of enhancing recharge of the aquifer. It means increasing public awareness on water conservation and not allowing development in areas serviced by public water. It means building treatment plants fed by river water (Saigon river, Ban Nigh river) and its input into distribution networks (e.g., municipal wells). This is costly but can bring extraction from an aquifer to volumes equalled by recharge and thus arrest subsidence.

To reduce the effects of flooding, sewage and drainage works, including canals, have to be maintained and kept clear of debris accumulations.

Pumping stations should be installed where necessary to move floodwaters to the sea as in New Orleans and Miami, USA. Protection from encroachment and flooding of low coastal areas by high and spring tides could be done with dikes or seawalls, but as with other fixes this would require investment to install and maintain the defenses. The costs have to be considered in terms of what yearly costs have been (millions of dollars) to repair flood damaged structures and infrastructure and what will be lost to the city (functionality) and to the nation (productivity) if necessary actions are not taken with utmost urgency to secure a safe and productive future.

5.4   Loading Greenhouse Gases into the Atmosphere

The blanket of greenhouse gases in the atmosphere is in great part humans' contribution to global warming and resulting climate change. The "blanket" includes carbon dioxide, methane, chlorofluorocarbons (in theory no longer emitted), nitrous oxide, and aerosols. This contribution is not abating as a principal measure of these gases shows with CO_2 continuing to increase in concentration. For example, from the 280 ppm atmospheric content at the beginning of the industry revolution in the 1880s, CO_2 content rose to 315 ppm in 1950, an increase of 35 ppm in 70 years. During the following 68 years, it rose to 386 ppm in 2010 to 400 ppm in 2015 and to 410 ppm in 2018, as measured by the Scripps Institute of Oceanography Mauna Loa Network. The high rate of increase post 1950 of 95 ppm is attributed to reconstruction and growing industrialization after WWII.

The human activities that cause the steady increase in greenhouse gases in the atmosphere have been discussed earlier: emissions from industrial development especially from the burning of fossil fuels (coal, fuel oil, natural gas, gasoline, and diesel fuel) in power plants and other industries and from internal combustion engines that power vehicles. It should be noted that there are natural additions of greenhouse gases to the atmosphere such as emissions from volcanoes especially during eruptive phases. The problem is further exacerbated by humans cutting down of great swaths of forests (e.g., in Brazil and Southeast Asia) that remove sinks for CO_2, thereby adding to the increasing mass of this gas in the atmosphere.

5.4.1 Profile of Ocean Warming and Projected Effects

In 2013, the Intergovernmental Panel on Climate Change reported a warming of oceans of 0.44 °C (0.79 °F) in the upper 75 m (42 fathoms) and to a lesser degree to 700 m (393 fathoms), as measured from 1971 to 2010 [18]. The warming or increase in the oceans' heat contents over time corresponds well at the 95% confidence level with the increase in atmospheric CO_2 contents [19]. The warming lessens the oceans capacity to act as sinks for CO_2 thus abetting its increase in the atmosphere. However, the oceans' heat content data given for the upper 2000 m of the oceans by the Intergovernmental Panel on Climate Change (in Watts per square meter [W/m^2]) averaged over the Earth's surface were smaller for that period (0.20–0.32 W/m^2) than higher value ones independently calculated with corrected data input to three climate models in the Coupled Model Inter-comparison Project 5 (0.36 ± 0.05, 0.37 ± 0.04, 0.39 ± 0.09 W/m^2). The models also indicated that post 1991, with additional data from 2005 to 2017, the rate of ocean warming as determined by the three computer models accelerated to 0.54–0.68 W/m^2 and the oceans' upper 700 m of water warmed at a greater rate than the waters to 2000 m. This warming corre-lates at the 95% confidence interval with the increase in the atmosphere CO_2 con-centration as observed at the Mauna Loa Observatory [20].

On the basis of a 40% increase in ocean heat content vs. earlier estimates, researchers projected that by 2100, the sea level rise from ocean warming alone (thermal expansion) could be 30 cm (12 in.) exclusive of sea level rise from melting alpine glaciers, ice caps, and ice sheets [20]. If there are inconsistencies in how data are used in computer analyses or new data are used as input, and a calculated increase in ocean heat content decreases to 10% or increases to 50% instead of 40%, a sea level rise would be 7.5 cm (3 in.) or 15 cm (6 in.), respectively. What some may consider a positive aspect is that ~93% of the extra heat trapped by increasing greenhouses gases in the atmosphere warms the oceans and does not abet increases in land temperatures.

If the conclusions of researchers cited above are further confirmed, this increases the threats that coastal cities have to adapt to and prepare for. The warmer oceans evaporate more water into clouds that then precipitate more rain when over oceans and from tropical storms and hurricanes (typhoons, monsoons) over land. The warmer water acts as fuel for such storms as they move over it increasing energy and force as they approach land, make landfall, and hit the coast (and cities) with power-ful winds that drive seawater surges inland and precipitate heavy rains that can cause floods when they are sustained for a period of time or released as torrential downpours. Such conditions have been deadly for coastal cities in developed, devel-oping, and less developed nations. The enhanced rainfall is a stimulus for landslides in hilly terrain underlain by sedimentary rocks with an attitude that dips or angles at as little as 15° to the face of a slope and makes them susceptible to downward movement by the pull of gravity. The added weight of water and lubrication of water along sedimentary rock contacts decrease the rock resistance and can ultimately result in a landslide. Landslides are not generally a significant problem in coastal cities.

5.4.2 Added Coastal City Problems Driven by Global Warming: Drought, Fisheries, Heat Waves, Wildfires

Climate change driven by global warming can cause additional significant direct and indirect problems for populations in coastal cities. If these are projected to continue and increase in future years then it is necessary to move into an adaptation mode now. An extended drought, one that lasts for years, is prominent among these and seems to be happening more frequently in many parts of the world. Coastal cities can adapt to a threat of future droughts that draw down their water reserves (e.g., in reservoirs) to dangerously low levels by building seawater or brackish water desalination plants that release cleansed water into an existing water grid during a drought crisis.

Southeast Australia suffered drought (low rain) conditions from 1997 to 2009 that affected its coastal cities and great agricultural regions. It was named the Millennium Drought. In addition to the direct stress on the domestic water supply for drinking, cooking, and personal hygiene in South/Southeast Australian coastal cities, there was an economic stress on populations from higher prices for vegetables, fruits, and nuts because of lack of sufficient water for irrigation and loss of field, tree and bush crops. Beef prices rose as well because of diminished feed crops and loss of grazing nutrition so that supplementary feed had to be bought and water trucked in to keep cattle, dairy cows, and sheep healthy. For some farms the lack of water was so bad that herds were moved to where there was better food/water availability. Cities responded by following stages of an increase in water use restrictions as reservoir capacities fell. In Melbourne, a city of 5 million inhabitants, the capacity fell to 26.7%. The city imported water via pipeline from a river and began construction of a desalination plant with funds from the government of Victoria state. It was completed in 2011, post-drought, with criticism from politicians about the funds expended. However, when another short drought period ensued in 2012, the plant delivered water to make up for shortfalls and that muted the critiques. Victoria state also used recycled water for irrigation, toilets, and clothes washing where there were secondary pipes to move the water [21]. Another important result of drought in Australia was that hydroelectric dams produced less electricity because of a reduced river flow especially in 2007 after the driest year on record (2006) so that restrictions on electricity use came into play. Output from coal-fired plants was cut back as well because of lack of fresh water for cooling needs.

Other major Australian coastal cities also constructed desalination plants. For example, Sydney with a population of 5.1 million people saw reservoir capacity fall to 33% and built a desalination plant that opened in 2009. Adelaide hastened to build a desalination plant that will supply 50% of the water needs for its 1.3 million citizens. Brisbane used pipeline water from rivers and also had treatment plants that processed gray waters to help meet the needs of its 1.1 million people [22]. Western Australia suffers drought conditions as well. Perth, with 2 million citizens, is perhaps the richest state with mining and major industries. This coastal city worked to drought-proof itself with two seawater desalination plants south of the city that provide almost half of its water needs, and by treating wastewater that releases clean water into the water distribution network [23].

Cape Town, South Africa, a coastal city of 4 million people, suffered a severe 3 year drought from 2015 to 2018. Seventy percent of the water use is in homes. Emergency measures were taken to reduce water use from the city's reservoir as its capacity fell. Zero dates were predicted for when the reservoir capacity fell to 13.5% and taps would be turned off. However, as citizens conserved water use, the zero dates were extended as allowances were stepwise reduced to 50 L (13 gallons) per person daily (vs. 185 L [49 gallons] per person daily global average). This meant flushing toilets once a day and limiting showers to 10 L (~2.6 gallons). Before zero day arrived, the drought broke and by June, the reservoir reached 43% capacity and by September 70% capacity. The economy suffered from the drought as wine production, an important export commodity, dropped 20% and vegetable and fruit production dropped 15% annually. Cape Town is planning the construction of a seawater desalination plant and a search for groundwater resources [24].

No continents save Antarctica have coastal cities immune from drought. Southern California is a dry region with coastal cities (e.g., Los Angeles, San Diego) that draw water from the Colorado River and from Northern California rivers. The area has seven desalination plants with two more under construction to provide extra water as needed during periods of drought.

Global warming can affect the economy of coastal cities that have an important fishery sector. The warming ocean waters to 75 m depth (42 fathoms), as cited earlier, has as a consequence fish that could not adapt to the warming migrated to cooler waters, to the north in the northern hemisphere and to the south in the southern hemisphere. Fishing boats have to venture farther to harvest the fish using more time and fuel thus reducing income from their catches. This migration could become an international problem. Products from fisheries in Peruvian waters bring in important foreign exchange. If an important mass of fish from warming waters off Peru migrate to cooler Chilean waters, Peru would lose income that Chile would gain. Given that this is driven by global human actions, a moral question is whether there should be a sharing of a catch of "Peruvian" fish in Chilean waters.

A heat wave often accompanies drought in coastal (and inland) cities. If one is long lasting for days or weeks, citizenry has to be protected.

Adaptation to this natural hazard can be attained by first having cooling stations available for mobile citizens. Second is to have a plan and infrastructure to bring fans and supplies to the elderly and homebound during a heat wave. South and Southeast Australian and Southern Californian coastal cities suffer through such events.

A combination of drought and heat wave can set the stage for wind driven destructive wildfires that can affect coastal cities such as those that seasonally take place in South and Southeast Australia and Southern and Northern California. An adaptation already in place where wildfires are a seasonal threat to people, property, and economics is to have fire watchers that can spot the beginnings of a conflagration and route fire fighting teams to limit a fire's advance. This is easier said than done depending on weather/wind conditions, topography, and accessibility by fire fighters and equipment. Continuous educational campaigns on how to prevent wildfires is in place in regions affected by fires. Homeowners with a large enough plot can respond to the threat of wildfires by reducing their exposure to advancing fires by clearing drought/heat dried vegetation (fuel) around a property for a perimeter of 10 m (33 ft) if possible, and when building or rebuilding, using fire-resistant exterior materials instead of wood or asphalt roofing materials.

Finally, it must be noted that in addition to heat waves and their effects on land, there are marine heat waves in the oceans. These occur when seawater has a temperature much warmer for at least 5 consecutive days than the 90th percentile value based on a 30-year historical baseline. Researchers report that analysis of almost a century of data (1925–2016) indicate that marine heat waves were more frequent by 34% and lasted longer by 17% and that there has been an increase of more than half in yearly heat wave days globally. This tends to follow the increases in mean ocean temperatures cited earlier in this chapter [25]. These >5 day periods have disrupted ecosystems including neritic zones in coastal regions. The ecosystems include foundation species (corals [reefs]), seagrass beds, and kelp forests that protect biodiversity by providing shelter for fish (and spawning habitats) from predators, and by serving as a food source. The question that has to be answered is whether an ecosystem impacted by a period or repeated periods of marine heat waves can recover once the seawater temperature "normalizes" or if marine heat wave waters increase in temperature as global warming continues. Both conditions will disrupt populations dependent on commercial fisheries and aquaculture that are important to food security and economies of coastal communities as the previously noted migration of fish [26].

References

1. U. S. Geological Survey. What is the probability that an earthquake will occur in the Los Angeles Area? In the San Francisco Bay Area? http://www.usgs.gov/faqs/what-probability-earthquake-will-occur-los-angeles-area-san-francisco-bay-area
2. Teshigawara M (2012) Appendix A: outline of earthquake provisions in the Japanese Building Codes. In: Preliminary Reconnaissance Report of the 2011 Tohoku Chica Taiheigo Oko Earthquake, Geotechnical, Geological, and Earthquake Engineering, vol 23. Springer, Tokyo, pp 421–446
3. California Building Standards Commission (2013) California existing building code, vol 1 and 2. California Building Standards Commission, Sacramento, 752p
4. Siegel FR (2016) Mitigation of dangers from natural and anthropogenic hazards: prediction, prevention, preparedness. Springer Briefs in Environmental Sciences, 127p

5. Wilcox K (2015) New bridge unfolds in Japan. ASCE Magazine. http://www.asce.org/magazine/20150818-new-bridge-unfolds-in-japan/
6. FEMA (1993) Against the wind: protecting your home from hurricane wind damage, 247, ARC 5023, 6p. http://www.fema.gov/media.library-date/20130726-1505
7. FEMA (2010) P-804 Wind retrofit guide for residential buildings. http://www.fema.gov/media.library-date/20130726-175
8. Xu Y-S, Shen S-L, Ren D-J, Wu H-N (2016) Analysis of factors in land subsidence in Shanghai: a view based on strategic environmental assessment. Sustainability 8:573–584. https://doi.org/10.3390/su8060573
9. Siegel FR (2018) Cities and mega-cities. Problems and solution strategies. Springer Briefs in Geography, New York, 117p
10. Caen M (2018) New project aims to reinforce Millennium Tower. CBS, San Francisco
11. Osathanon P (2015) Action required to stop sinking the capital. The Nation, Thailand Portal
12. Lorphensri O, Ladawadee A, Dhammasarn S (2011) Review of groundwater management and land subsidence in Bangkok, Thailand. In: Taniguchi M (ed) Groundwater and subsurface environments, part 2, chap 7. Springer, Tokyo, pp 127–142
13. Aobpaet A, Cuenca MC, Hooper A, Trisirisatayawong I (2013) INSAR time-series analysis of land subsidence in Bangkok, Thailand. Int J Remote Sens 34:2969–2982
14. Ishitsuka K, Fukushima Y, Tsuji T, Yamada Y, Matsuoka T, Giao PH (2014) Natural surface rebound of the Bangkok plain and aqua characterization by persistent scatter interferometry. Geochemistry, Geophysics. Geosystems 15:965–974. https://doi.org/10.1002/2013GC005154
15. Tran Ngoc TD, Perset M, Strady E, Phan TSH, Vachaud G, Quertamp F, Gratiot N (2016) Ho Chi Minh City growing with water-related challenges. In: Water, megacities and global change, 27p. http://www.eaumega.org/wp-content/uploads/2016/05/HCMC-MonographyEN.df
16. Nguyen QT (2016) Therein causes land subsidence in Ho Chi Minh City. Procedia Eng 142:334–341
17. Wikipedia. List of tallest buildings in Vietnam. http://www.en.wikipedia.org/wiki/List_of_tallest_bilngs_in_Vietnam. Accessed 2018
18. Intergovernmental Panel on Climate Change (2013) Fifth assessment report. In: Climate change 2013. The physical science basis. Cambridge University Press, Cambridge, pp 215–315, 29p. http://www.ipcc.ch/report/ar5/wgl/
19. Cheng L, Trenberth EE, Fasullo J, Abraham TP, Boyer K, Zhu J (2017) Taking the pulse of the planet. EOS, 98. https://www.eos.org/opinions/taking-the-pulse-of-the-planet
20. Cheng L, Abraham J, Hausfather Z, Trenberth KE (2019) How fast are the oceans warming? Science 363:128–129. https://doi.org/10.1126/science.aav7619
21. Gray S (2016) Melbourne's desalination plant is just one part of rough-proofing water supply. The Conversation, 3p
22. Radcliffe JC (2015) Water recycling in Australia – during and after the drought. Environ Sci Water Res Technol 1:554–562. http://www.xlink.rsc.org/?DOI=CSEW00048C
23. Morgan R (2015) 'Drought-proofing' Perth: the long view of Western Australia water. The Conversation. http://www.theconversation.com/drought-proofing-perth-the-long-view
24. Wikipedia (2018) Cape Town water crisis. http://www.en.wikipedia.org/wiki/Cape_Town_water_crisis
25. Oliver ECJ, Donat MG, Burrows MT, Moore PJ, Smale DA, Alexander LV, Benthuysen JA, Feng M, Gupta AS, Hobday AJ, Holbrook NJ, Perkins-Kirkpatrick SE, Scannell HA, Straub SC, Wernberg T (2018) Longer and more frequent marine heatwaves over the past century. Nat Commun 9:1324–1336. https://doi.org/10.1038/s41467-018-03732-9
26. Smale DA, Wernberg T, Oliver EJC, Thomsen MS, Harvey BP Straub SC, Burrows MT, Alexander LV, Benthuysen JA, Donat MG, Feng M, Hobday AJ, Holbrook NJ, Perkins-Kirkpatrick SE, Scannell HA, Gupta AS, Payne B, Moore PJ (2019) Marine heatwaves threaten global biodiversity and the provision of ecosystem services. Nat Clim Chang, Letter. https://doi.org/10.1038/s41558-019-0412-1

Chapter 6
Disease Protection in Sea Coast (and Inland) Cities: Problems in Dense Populations with Shantytowns/Slums

6.1 Introduction

As discussed earlier in this book, there are sea coast cities worldwide that are at risk from floods, storm surges, and extreme weather conditions such as wind-driven high category hurricanes (typhoons, monsoons), or drought and heat waves, plus along Pacific Ocean coasts, earthquakes and tsunamis. A municipality and its public health services have to be prepared to adapt to their preparations to deal with what contemporary experiences and history reveal are the most likely physical hazards and diseases to impact it. The two main preparatives are first to be able to care for the injured during a hazard event at well-staffed and supplied hospitals and medical clinics or field hospitals. The second is to fill the basic needs of affected populations with clean water, food, shelter, toilets, waste collection, and if necessary power restoration. These primary responses will help to reduce the chance of an onset and spread of disease. It is important to activate search and rescue teams help citizens isolated or trapped by the event that did not or could not evacuate such as in the case of extreme weather and flooding. In addition, where there have been deaths, bodies should be recovered and interred as soon as possible in order to prevent sickness that might originate from them. For instances of collapsed structures with people trapped inside, equipment that can move debris and rescue people or recover the deceased is necessary as would be the case for less common strong earthquakes as well as some extreme storm events. What was just described may be a good template to follow, but in reality, many sea coast cities in developing and less developed countries do not have the resources to cope with severe hazards that might impact them. Here, the World Health Organization and developed nations, large and small, have sent in personnel, supplies, and equipment to help an impacted sea coast city (or inland city) in the past and will surely continue to do so in the future.

F. R. Siegel, *Adaptations of Coastal Cities to Global Warming, Sea Level Rise, Climate Change and Endemic Hazards*, SpringerBriefs in Environmental Science, https://doi.org/10.1007/978-3-030-22669-5_6

6.1.1   Disease Control

Disease control after a natural disaster often means initially dealing with diarrheal diseases (dysentery) from the ingestion of contaminated water or food, and lack of sanitation facilities (toilets, hand-washing stations). It may mean dealing with cases of the infectious diseases like cholera (bacterium *Vibrio cholera*) or typhoid (bacterium *Salmonella typhi*) that originate from contaminated water or food or from contacts with cholera or typhoid carriers. To prevent a spreading epidemic status means making safe water and food, sanitation and hand-washing posts available. In the case of cholera, an epidemic may be avoided by isolating the cholera carriers and treating them (with hydration and antibiotics) or oral cholera vaccines where the disease is endemic until the threat of spread is under control. During March 15–16, 2019, the monsoon Idai flood ravaged Beira, Mozambique, a sea coast city of 533,000 people, and tracked inland into areas of Zimbabwe and Malawi flooding them and killing a total of over 1000 people. In Beira, the drinking water supply and the sewage infrastructure were damaged. Cholera cases were diagnosed and the disease spread rapidly with over 4000 cases identified with 7 deaths by the April 9. The WHO assembled more 975,000 doses of cholera vaccine (oral) and starting April 3 delivered more than 487,000 initial doses to Beira citizens and those in other affected populations, but only about 2 weeks after the disease was first diagnosed. A second inoculation 2 weeks later gives full protection from the disease. Adaptation would require vaccine storage stations set up in regions where outbreaks of endemic and other infectious diseases occur so that those that can be controlled by vaccination could be serviced more rapidly than was the case in Mozambique.

6.2   Infectious Diseases

In the above section, infectious diseases that could follow natural disasters that affect sea coast cities were discussed. In this section, infectious diseases that can affect sea coast cities by developing from within or those that are carried in from outside sources are considered. These would be water- and food-borne contaminants and vector-borne bacteria/viruses (by insects, animals, humans). Important factors that have to be considered by public health personnel in sea coast cities in order to be prepared to deal with disease include a location's latitude and elevation as they influence climate (temperature and humidity).

6.2.1   Sea Coast City Climate Zones and Endemic Diseases

Important sea coast cities are located in most geographic climatic zones. As such, and based on a history of urban health problems, their populations may be susceptible to one or more than one infectious disease that public health services should be prepared to treat. Six principal climate zones (Mediterranean, tropical, arid, temperate, polar, and mountainous) are classified on the basis of average monthly temperature and precipitation. Temperature and precipitation are determined in grand part by latitude, altitude/landform, and the influence of surface and subsurface ocean currents flow paths, location of mountains and their elevations, and seasonal wind directions. The principal zones are a simplification of a 12 zone classification, each with sub-zones (Table 6.1) [1]. Missing from this classification is a mountainous one because with their masses and elevations they establish their own micro-zones and affect weather patterns of some of the other zones.

Sea coast cities in a similar stage of economic development and in the same climate zones are likely to have exposure to a similar group of endemic infectious diseases. For example, in Southeast Asia, Jakarta, Bangkok, and Ho Chi Minh City, are hot (30 °C) with higher humidity (75% RH) and have endemic infectious diseases that originate from contaminated water and food, and those that originate from vector-borne carriers. Country reports show that in the endemic infectious group, all live with the threat of bacterial diarrheal diseases, and two cities live with the threat of Hepatitis A and typhoid fever. In the second group, the three cities have

Table 6.1 Twelve zone, multi-sub-zone climate classification [1]

Rainforest	High rainfall (175–200 cm annually, mean monthly temperature >18 °C)
Monsoon, typhoon, hurricane	Seasonal regional rainy season (North and South America, Sub-Saharan Africa, Australia, and East Asia)
Tropical savanna	Semi-arid to semi-humid regions in sub-tropical and tropical latitudes with temperature at or >18 °C with rainfall between 75 and 127 cm annually (widespread in Africa, found in India, and northern parts of South America, Malaysia, and Australia)
Humid sub-tropical	Winter rainfall mainly from thunderstorms. Mainly on east side of continents between about 20° and 40° latitudes away from the equator
Humid continental	Large seasonal temperature range with 3 months of temperatures >10 °C and coldest month temperature <−3 °C
Oceanic climate	Along west coasts of all continents at middle altitudes and in SE Australia with lot of precipitation year-round
Mediterranean climate	Hot, dry summers and cold, wet winters. Mediterranean Basin countries, western North America, parts of Western South Australia, SW South Africa
Desert	Little precipitation (dry), large temperature range daily (up to 45 °C and down to 0 °C) and seasonally hot, during day, cold during night, all year except Antarctica
Steppe	Dry with temperature that can range from 40 °C during the summer to −40 °C during the winter
Subarctic climate	Little precipitation and with permafrost. Temperatures of >10 °C for 1–3 months and for 6 winter months <0 °C
Tundra	In the far Northern Hemisphere especially in extensive areas of northern Russia and Canada. Very cold all year
Polar icecap	Dry, very cold, all year

exposure to dengue fever and malaria while two are threatened by Japanese encephalitis, and one with HIV/AIDS and tuberculosis.

In Africa, Mombasa, Kenya in East Africa, population 537,000 and Luanda, Angola in West Africa, population 2.5 million, are located relatively close to the equator (4° 3′S and 8° 50′S, respectively) and have similar temperatures of 25–3 °C and 60–65% relative humidity defining their Mediterranean/sub-tropical to hot humid tropical climates. However, unlike the Southeast Asian sea coast cities with similar infectious disease burdens cited above, Mombasa and Luanda infectious disease burdens differ greatly. Mombasa residents are at risk from malaria, tuberculosis, HIV/AIDs and recently from a resurgent occurrence of polio, measles, and kanazaar (parasitic disease *Leishmania donovani*). Luanda, Angola residents are at risk from yellow fever, malaria, typhoid fever, and Hepatitis A. Mombasa, an important Kenyan port city also serves Uganda, Rwanda, the Eastern region of the DR Congo. This put the Mombasa population at risk of infectious diseases that may be carried from those countries by people or products. This possibility for infectious disease transfer was supported when public health personnel recently identified an Ebola strain in Kenyans who visited West Africa. The West African region had suffered an Ebola epidemic during 2013–2016. The disease likely began in a village of southeast Guinea as a young boy infected by bats (bushmeat) was carried by exposed villagers to its capital and port city Conakry (population 1.8+ million). The Ebola virus was spread by carriers to the Liberian capital and port city Monrovia (population 1 million with 41 doctors), and to the Sierra Leone capital and port city Freetown (population 1+ million). The World Health Organization, medical teams and NGOs from other nations worked to stop the epidemic but only after more than 11,300 died and more than 28,000 were infected and saved. An as-yet unapproved Ebola vaccine was used here and was instrumental in halting the spread of the Ebola virus.

Major sea coast cities in South America and the Indian sub-continent have similarities and differences of infectious disease risks as do the sea coast cities Mombasa and Luanda cited above. Public health services staffed with appropriate personnel, supported financially, with medicine stocks plus WHO vaccination stores can be prepared to use them to halt the spread of an infectious disease to prevent an epidemic or to deal with an epidemic that may develop. On August 1, 2018, there was an outbreak of Zaire Ebola virus in North Kivu province of the DR Congo that spread to Ituri Province. In March, 2019, WHO reported that there were more than 1029 cases and 642 deaths. The public health ministry medical experts identified the disease, did contact tracing, contained and treated infected people, and worked to educate the threatened populations of the risk of not reporting Ebola cases, and unsafe burial practices (touching and washing the dead) that is meeting cultural resistance. As important in dealing with the epidemic was the acceptance of the use of the experimental Ebola vaccine that helped stop the spread of the Ebola disease in West Africa a few years earlier. The vaccine was given to 3300 people who had direct or indirect contact with people with the disease. However, the application of the Ebola vaccine and a halt in the rise of infections has been hampered by armed group violence and unsafe burial practices. The disease is recurrent in the DR Congo with ten outbreaks since the discovery of the disease in 1976 [2]. The hope is that it will not become entrenched.

One adaptation to infectious disease prevention in this time of ready transnational transport of people globally via sea, air, and land is vaccination control. Vaccinations against infectious diseases keep a nation healthy and productive. A low rate of vaccination in a country or regions for years on end can be damning to its citizens and to people where they travel within that country or internationally. The WHO warns that vaccine hesitancy is one of the top ten global threats of 2019. Vaccination hesitancy can be the result of personal philosophical or religious beliefs that allow parents to opt out of this protection for their children as in the United States where there have been major measles outbreaks in eight states and reported cases in 20 other states. In many of these states legislatures are proposing laws to minimize exemptions to measles vaccination requirements for children of school age. A judge in Brooklyn, New York, following a serious outbreak of measles in children of a religious sect during April, 2019, mandated application of the MMR measles vaccine (including mumps and rubella) to all their children to protect the general public. This was done over the religion leaders objections to vaccinations.

During 2018, Europe (53 countries) reported that measles killed 72 children and adults of the 82,500 people that contracted the measles virus due in part to a vaccination rate (<90%) less than the 95% WHO deems necessary to prevent epidemics and in part to easily travel between countries. In a recent deadly outbreak, Madagascar is working to contain a measles epidemic that began in September, 2018 and by February, 2019 had killed more than 900 people (64% children to age 14) and infected more than 68,000 persons. This epidemic is the result of a low immunization rate for many years and has spread throughout the country including Madagascar's most important seaport, Toamasina, and other ports of entry. This, in spite of the fact that a safe and effective vaccine, has been available for the past 50 years but whose application has not been mandated by the national government. Travelers from Madagascar to other countries are potential carriers for about 2 weeks (4 days before a rash appears, plus 6 days during the rash, and for 4 days after the rash disappears).

For this reason, the vaccination history of people entering a sea coast city or other ports of entry who come from or have passed through areas that are suffering or have suffered infectious disease events is important. This would require the collaboration of immigration officials who would be charged with checking the vaccination cards of people (ship personnel, tourists, immigrants, and asylum seekers) at sea coast city ports of entry. The same would be true for other ports of entry. It may require making spot checks of, for example, fever or an observational feature (sweating, paleness) of arriving passengers from specific locations suffering from outbreaks of infectious diseases. This was the case by fever checks in airports and other points of national entry after China revealed the spread of the SARS (severe acute, respiratory syndrome) infectious disease during 2002–2003.

6.3   Human Activities Sources of Non-Communicable Diseases

Non-communicable diseases whether in highly and densely populated sea coast or inland cities can sicken and kill citizens. The focus on the causes of non-communicable diseases (NCDs) that are responsible for 41 million of premature deaths annually and a far greater number of debilitating sicknesses has been on behavioral and metabolic causes and how to reduce their threats to societal health and its effect on city and national economies. These causes include tobacco use, alcohol abuse, poor nutrition, lack of exercise, obesity, and hypertension. Of the 41 million premature deaths cited above, ~35 million are in low and middle income countries. Seventy-nine percent of the NCDs are attributed to cardiovascular disease (17.9 million), cancer (9 million), respiratory disease (3.9) million, and diabetes (1.5 million). Public health education programs have to be continually implemented to reduce harmful behavior and causes [3].

6.4   Pollution

To this point in time, not enough attention has been given to pollution as a principal factor in NCDs illnesses and even premature deaths and their socio-economic consequences [4]. These are often expressed by a calculation of DALYS (disability adjusted life years), the total number of years lost to illnesses, disability, or premature deaths in a given population for a certain disease or disorder [5]. This can be a serious problem in high population sea coast and inland cities with high densities of citizens especially when there are edge neighborhoods with poor and marginalized people living in shantytowns/slums with unsafe water, lack of toilets, and problems with garbage/waste collection. The loss of productivity because of DALYS lowers GDP levels by up to 2% annually, especially in low middle income countries (incomes of US$1026 to US$4035), such as Bangladesh, Egypt, India, Indonesia, Pakistan, and Vietnam. Cities are most affected because they account for 85% of global economic business [6].

Pollution is a bane of humanity. Air, water, soil, and food pollutants contribute to the premature deaths of millions citizens and sickens many millions more from NCDs. Pollution-caused diseases accounted for about 9 million premature deaths in 2015 or 16% of all deaths globally (>56 million). Of the 9 million premature deaths, about 70% of these (e.g., cardiovascular disease, COPD) were from pollution-caused NCDs, mainly in low income and low middle income countries [4].

Pollution is most damaging to people and economies where environmental laws protecting people and ecosystems from pollution have not been instituted or if laws have been passed but are not enforced. Pollution does not distinguish between sea coast and inland cities but as noted above is most damaging where there are high and dense populations breathing, drinking, and eating pollutant bearing matter. The effect of ingestion of pollutants that cause NCDs is not instantaneous but acts slowly over a period of time gradually attacking the normal functioning of body organs as pollutants access and bioaccumulate in them. Pollution can be greatly mitigated in three ways: (1) taxation has been effective in minimizing/eliminating the mass of pollutants emitted into the atmosphere or discharged into waterways or onto soils; (2) investment in pollutant capture and control equipment and safe disposal of the captured pollutants; and (3) legislation enforced by the threat and actions of closure of a pollutant-generating operation, by fines, or by incarceration of responsible individuals.

Of the ~9 million premature deaths caused by pollution, ~7 million are from indoor air pollution (2.9 million) and outdoor air pollution (4.2 million) [7]. There are 1.8 million people that die annually from water pollution (bacterial, viral, heavy metals) and 0.5 million from soils polluted with heavy metals and toxic chemicals (e.g., pesticides) ingested through foods. From an economic perspective, pollution frequency costs in the range of 4–5% of a nation's GDP and for some this exceeds what they receive in development aid [8].

6.4.1 Air Pollution

Diseases that originate from indoor air pollution include stroke, coronary artery and ischemic heart diseases, chronic obstruction pulmonary disease (COPD), acute lower respiratory infections in children, and lung cancer. Outdoor air pollution diseases are much the same but with less coronary artery disease.

Sickness and deaths from indoor air pollution are the result of poor or no venting of gases and fine size particulates (<2.5–10 μm) from the fuel used (e.g., soft coal, charcoal, wood, dried animal waste) for cooking and/or heating. Improved venting of indoor generated pollutants to the outside can reduce sickness and deaths especially for the more susceptible old people and small children but will add these toxins to outdoor air pollution. The effects of indoor air pollution can be eliminated by bringing natural gas to homes for cooking and heating and proper venting of emissions. This solution is lacking in some major sea coast and inland cities with highly populated shantytown/slum in inner city or edge neighborhoods.

Outdoor air pollution comes from multiple sources such as coal-burning power plants, various industries, vehicular traffic, and construction. These sources, individually or in consort, emit or raise fine size particulates (<2.5–10 μm), and other toxic emissions into the air that include sulfur dioxide, nitrous oxide, and heavy metals (e.g., lead [Pb], mercury [Hg], arsenic [As], cadmium [Cd], and others). These toxic emissions may be inhaled or deposited onto soils or into waters where they can enter a food chain via agricultural crops (e.g., Cd in rice) and food fish (e.g., Hg in tuna, swordfish, king mackerel). Under some meteorological conditions, toxic emissions from industries and vehicular emissions will contribute to a sickening or killer smog. Air pollutant abetted diseases are most prevalent among major sea coast and inland cities with upwind nearby major industrial development and in shantytown/slum populations vs. economically more advantaged populations. The mandated use of pollutant capture and control technologies can mitigate the emissions of toxins from industrial sources cited and together with the inclusion of catalytic converters on vehicles can reduce the number of premature deaths abetted by outdoor air pollution. As cited previously, taxation of emission masses can mitigate the problem as can fines, plant closure until facilities becomes compliant with government legislative mandates, or the threat of imprisonment of plant owners [9].

Several major sea coast cities with high populations with very high and relatively high densities in developing countries suffer greatly from air pollution. These include Dhaka (47,400 people/km²), Mumbai (26,400/km²), Abidjan (13,700/km²), Kolkata (11,200/km²), Jakarta (9800/km²), Ho Chi Minh City (6800/km²), Bangkok (5200/km²), Guangzhou (5200/km²), Zhangjiang (5000/km²), Tianjin (4900/km²), and Ningbo (4000/km²). In some of these cities, the air pollution threat is elevated by high numbers of poor populations living in shantytowns/slums where densities are higher than the average for a city and where the health status of a population is not as resistant to sickness and disease as it is in economically advantaged inner city populations. For example, in 6.5 million people of the Mumbai 23.3 million population live in shantytowns/slums as do 3.8 billion people of Kolkata's 13.3 million population. Similarly, of the 17.4 million population in Dhaka, 5.2 million live in shantytowns/slums. Perhaps most striking is that of the Lagos population of 14.8 million people, 8.8 million live in shantytowns/slums. In comparison to the densities of major coastal seaports, those in developed countries are generally less and without a significant slum populations (e.g., Rotterdam [3100/km²], Hamburg [2700/km²], Los Angeles [2300/km²], New York [1700/km²], and Antwerp [1600/km²]) [10].

Because of the concern for citizens health and the economic losses from workdays lost or even years of an individual's contribution to a nation's development lost, municipal governments have been monitoring air pollution and identifying specific sources of air-borne toxins. In some cases, they are enforcing existing legislation or carefully planning legislation with enforcement powers that mandate the use of best available technologies at all toxic emissions sources and safe disposal of the captured pollutants. If enforcement is applied, it will greatly reduce toxin loading of pollutants into the atmosphere and hence reduce the risk to the health and premature deaths of great numbers of people. This will be effective only if enforcement of laws curbing air pollution is not just an empty threat because of a nation's drive for continued economic development or because of a culture of corruption.

6.4.2   Water Pollution

Water pollution comes from the atmosphere as rain precipitates toxins from industrial emissions, especially coal-fired power plants and other sources into waterways or onto land where contaminants may run off into waterways. Water pollution comes as well from toxins in industrial effluents discharged onto land that can run-off into streams and rivers. Rainwater or streams that flow through talus piles of mine wastes can dissolve out heavy metals and carry them overland into waterways. Acid mine drainage from abandoned and functioning operations contaminates streams/rivers into which it discharges. Bacterial pollutants from human wastes where toilets are not used or where the wastes are collected but not treated can pollute waterways as do wastes and agricultural chemicals as runoff from farmland and commercial animal husbandry (e.g., cattle feed lots, chicken production). The toxin runoff, whatever the source(s), affects not only humans but ecosystems and their natural resources that are important to people. Contaminated waters can be cleansed to drinking standards by passing them through a treatment plant before distribution to consumers. Toxin-bearing surface runoff cited above can also seep through a surface cover into aquifers and contaminate groundwater. Under some conditions, groundwater can dissolve heavy metals that may be in aquifer rocks thus polluting water downflow from where the toxic metals originate. As with surface sources of water, polluted groundwater can be treated to remove a heavy metal pollutant or more than one, or other chemicals before being released for human use (drinking, cooking, personal hygiene) or for irrigation of food crops. The sickening of citizens in well-populated sea coast cities from ingestion of pollutants in unsafe drinking (and cooking) water is a grave concern especially in poor neighborhoods that lack access to clean water vs. economically advantaged neighborhoods that receive treated clean water [11].

6.4.3   Soil to Food Pollution

Sea coast and inland cities with large and growing populations (e.g., in the millions) obtain much of their foods from distant national and international sources. The reality is that some agricultural products (e.g., grains, vegetables, tobacco) may contain toxins taken up from soil or irrigation water polluted by human activities. These include heavy metals and pesticides/herbicides. If contaminated products are consumed over time, their toxin contents can build up in body organs (e.g., kidneys, lungs) and cause NCDs (see Sect. 6.3). Toxin build up can also disrupt the normal functioning of nervous and cardiovascular systems, and renewal of skeletal matter. The ingestion of human generated soil/food pollutants can cause NCDs [12–14].

In addition to soils being polluted by industrial and agricultural sources, soils can have natural high concentrations of potentially toxic metals. If a rock contains heavy metals (e.g., black shale) and a soil forms from it as it degrades and decomposes, contaminant metal(s) may be retained in the soil. One pathway to humans is ingestion of one or more of the potentially toxic metals taken up by agricultural crops (food or a non-edible product such as tobacco) being grown in a heavy metal polluted soil. Another pathway is from rainwater that seeps through a naturally contaminated soil that reacts with it to release a heavy metal that seeps through the soil into an underlying aquifer. Here, the path is toxic metal from rock to soil to agricultural products or drinking/cooking water that can do the same harm to the human body as noted in the previous paragraph by bioaccumulation in and damage to vital organs and also cause medical symptoms and the onset of a NCD. Examples of NCDs caused by heavy metal poisoning abound in the literature [12, 13].

Whether a NCD will develop from ingestion of a potently toxic metal depends on several factors. These are the dose absorbed by a body, the route and duration of exposure to chemical species that can cause chronic or acute illness, age, gender, genetics, and health conditions of exposed individuals. Perhaps most studied in recent years are cadmium (Cd), lead (Pb), arsenic (As), mercury (Hg), and chromium (Cr). There are natural sources of potentially toxic metals (e.g., rocks containing arsenic-bearing minerals, ores, volcanic gases, and particulates). However, great masses of the heavy metal pollutant load in the atmosphere, water bodies, and soils come from human sources. These include industrial operations emissions and effluents, especially from coal-fired power plants, mining and smelters, and factories that manufacture batteries, pharmaceuticals, fertilizer and biocides, paper, pulp, and wood preservatives, among others. Sources for each metal or a group of metals that are thought to be linked to an epidemiology-defined disease node can be traced by following geological/geochemical principles used in the search for mineral deposits. Governments can then take action to greatly reduce or eliminate the source(s).

The fundamental question exists as to whether a government should make regular spot analyses to determine if there is a risk of heavy metal poisoning from agricultural food stuffs. In theory, the answer is yes. Realistically, however, analyses would be done but only after a heavy metal was identified fortuitously or if public health professionals detected symptoms of heavy metal poisoning in a population.

6.5   Heavy Metal Pollutants Harmful to People and Ecosystems

Heavy metals are important to industrial development as well as to human health. There are heavy metals that are essential micronutrients for the optimal functioning of biological processes and organs in humans in a quantity of a few milligrams (mg) or micrograms (μg) daily. These are Fe (iron), Zn (zinc), Co (cobalt), As (arsenic), Cr (chromium), Cu (copper), Mn (manganese), Mo (molybdenum), Se (selenium), and V (vanadium). **However, these may become toxic if ingested and bioaccumulated over time to higher concentrations than are necessary to satisfy nutritional needs.** As noted in a previous paragraph, heavy metals that are most toxic to humans (+ animals and ecosystems) if ingested regularly in water and/or food are As, Pb, Hg, and Cd. These metals can bioaccumulate in a body and have toxic effects on different vital organs. In addition, they can interfere in the functioning of essential elements and their metabolic processes. These four heavy metals have been well-studied by toxicologists and other scientists for sources, mobility in ecosystems, and toxic effects. Their maximum allowable World Health Organization limits via oral intake in μg/day are Pb (10), Hg-inorganic (15), As-inorganic (15), and Cd (5) [15].

In the past and the present, old established sea coast cities have been centers of industrial development ofttimes within and/or upwind and upriver of cities. This, and increasing urbanization as populations grow, fueled increased utility and transportation needs. The release of heavy metals from industrialization, utility sources, and vehicles into a city environment and associated ecosystems on land and in the oceans can be an inherited legacy or existing danger to public health through bad air and contaminated water, soils, and foods. How sea coast cities (and inner ones) and national governments have adapted to mitigate toxic metal pollution is reflected in the health status of their populations.

6.5.1   Lead: Pb

Lead represents an inherited legacy that is an existing health threat in and surrounding old homes from Pb in drinking water that passes through utility and home pipes containing Pb or pipes joined with Pb-containing solder. This problem is being addressed by replacing the pipes, by regularly flushing and cleansing city water pipe networks, and by meeting WHO standards at treatment facilities. Lead bearing soil dust from vehicular sources has been basically eliminated since the phasing out of leaded gasoline that began in the 1970s and 1980s with the introduction of catalytic converters. However, inherited Pb in soils can be a health hazard. A city may opt to replace a soil, cover it, or otherwise remediate it especially at playgrounds. Cities were early in adapting to the Pb-based paint problem by 1965 recommendations that the amount of Pb in interior and exterior paint be reduced to <50% and subsequently to <0.1% in 1997. The UN and World Bank aim to eliminate its use in 2020. A reasonable recommendation to deal with the millions of homes with Pb-based paint is not to strip it off thus releasing toxic dust and chips to the air and surrounding soil but rather to repaint over it when the paint starts flaking. Lead reduction programs are essential to protect babies and small children who are especially threatened by Pb poisoning if they regularly ingest Pb through drinking water, hand to mouth transfer of contaminated soils, or chewing on flaking paint. Poisoning manifests itself by brain damage that can result in behavioral problems, learning deficits, and lowered IQ.

6.5.2   Cadmium: Cd

Cadmium is an existential health danger, a known carcinogen. A principal source of the toxin is in rice grown in Cd-rich soils or irrigated with Cd-rich water. A public outcry arose in 2013 in Guangzhou, China, a port city of 14.5 million people, when a report revealed that the rice tested in 8 of 18 samples from local markets had Cd concentrations above China's national food safety standards. Citizens were upset because of the knowledge that a long-term bioaccumulation of Cd in the body had been linked to chronic kidney disease/failure, the bone diseases osteoporosis and osteomalacia, diabetes, lung disease, impairment, cardiovascular disease, and cancer [15–17]. Rice is a staple food for close to half the world's population (~3.8 billion people) especially for poor populations in low and low middle income countries and its purity has to be assured. About 90% of the consumption is in Asia, with a rising demand in Africa and Latin America, **regions with a great number of commercially important sea coast cities.** Awareness of the potential problem has resulted in spot checks of rice crops to assure their purity. Scientists are studying how to reduce the bioavailability of Cd in a soil (paddy or not) to growing plants rice grains, by adjusting the pH of the growing medium [18, 19]. Geneticists that have improved micronutrient contents in rice crops and the plant tolerance to pests and herbicides are researching low Cd-accumulating rice cultivars growing in Cd-rich soils. These may reveal a gene that damps Cd uptake. Once identified, genetic engineers can modify receptive rice species that will greatly reduce uptake of soil Cd yet maintain their nutritional values. Cadmium can also access a body by inhaling cigarette smoke from tobacco grown in Cd-rich soils.

6.5.3   Arsenic: As

Arsenic has been shown to be the cause of chronic and NCDs in many countries with commercially important and high/densely populated port cities, especially in Bangladesh, India, Pakistan, Thailand, Vietnam, Taiwan, and China. Contaminated drinking water and cooking with untreated or insufficiency treated aquifer water is a principal source of the toxin. The As may develop within an aquifer as conditions change with water drawdown during growing season permitting oxidation and release from a mineral in aquifer rock into the water as either the oxidized chemical species arsenite (As^{3+}) or arsenate (As^{5+}). Both species can be toxic if enough is ingested but the (As^{3+}) form is 60 times more toxic [20]. The chronic response to As poisoning may be the appearance of lesions as pinkish to tan or skin colored papules and keratosis, an epidermal wart-like growth. Acute afflictions caused by the long-term As bioaccumulation are cancer of the lungs, liver, bladder, or skin [12, 15]. The World Health Organization warned that more than 200 million people are at risk of As poisoning from aquifer water. Water treatment facilities can eliminate this problem if installed, supplied with necessary cadres and chemical supplies, and maintained with a clean water distribution network that reaches all of a population.

6.5.4   Mercury: Hg

Mercury in food fish is an existing public health threat. Mercury emitted into the atmosphere from industrial sources, especially coal-fired power plants, rains into the oceans where microorganisms in the marine food chain convert it to the very toxic methyl mercury form (CH_3Hg^+) [21]. This chemical bioaccumulates in prey of large predator food fish (e.g., king mackerel, swordfish, tuna, tilefish, orange roughy, shark, marlin, and escolar) that further concentrate it to levels that can harm humans that regularly eat them. Shellfish bioaccumulate the toxin as well. Sea coast populations include much fish/shellfish in their diets, especially poorer citizens that count on fish/shellfish for 20% of their protein nourishment and 50% of their calorie intake. Mercury intake can be especially dangerous to pregnant women or women that breast feed their infants because the CH_3Hg^+ can pass to a growing fetus or an infant and cause neural damage. In adults, Hg poisoning can manifest itself as tremors, memory problems, depression, or diminished hearing and vision. Governments recommend how much Hg accumulating fish meat is safe to eat. For example, Canada recommends that pregnant women limit their intake to 150 g/month (5 1/4 oz), youngsters 5–11 years old eat no more than 125 g/month (~4 1/2 oz), and children 1–4 years old eat no more than 75 g/month (~2 1/2 oz). These values vary with the food fish eaten [22].

Lastly, it should be noted that there is a possible legacy source of heavy metal toxins or other pollutants: buried/hidden toxic wastes. These are disposal sites left by long gone industries in or near long established sea coast (and inner) cities. As urban centers grow to accommodate expanding populations (2.8 billion more people in 2050 than in 2019), perhaps 700 million added to port cities, areas to be urbanized should first be assessed to assure that no buried/hidden toxic waste disposal sites are present that could, over time, release toxins into an inhabited zone and pose a danger to people. If detected, such sites must be cleaned up before habitation is allowed.

References

1. Wikipedia. Köppen climate classification. www.wikipedia.org/wiki/Köppen_climate_classification
2. Claude KM, Underschultz J, Hawkes MT (2018) Ebola virus epidemic in war-torn estern DR Congo. Lancet 392:1399–1401. https://doi.org/10.1016/S0140-6736(18)32419-X
3. WHO (2018) Noncommunicable diseases. www.who.int
4. Fuller R, Rahona E, Fisher S, Caravanos J, Webb D, Kass D, Matte T, Landrigan PJ (2018) Pollution and non-communicable disease: time to end the neglect. Lancet 2(3):e96–e98
5. National Institute of MentalHealth. What are DALYS?. www.nimh.gov/dalys
6. Landrigan PJ, Fuller R, Acosta NJR, Nereus JR, Adeyl O, Arnold R, Basu NN, Baldé AB, Bertollini R, Bose-O'Reilly S, Boufford JI, Breysse PN, Chiles T, Mahidol C, Coll-Seck AM, Cropper ML, Fobil J, Fuster V, Greenstone M, Haines A, Hanrahan D, Hunter D, Khare M, Krupnick A, Lanphear B, Lohani B, Martin K, Mathiasen KV, McTeer MA, Murray CJL, Ndahimananjara JD, Perera F, Potočnik J, Preker AS, Ramesh J, Rockström J, Salinas C, Samson LD, Sandilya K, Sly PD, Smith KR, Steiner A, Stewart RB, Suk WA, van Schayck OCP, Yadama GN, Yumkella K, Zhong M (2018) The Lancet Commission on pollution and health. Lancet 391:462–512.
7. World Health Organization (2014) News release: 7 million premature deaths annually linked to air pollution. www.who.int/mediacentre/news/release/2014/air-pollution/en/

8. World Bank Group, Independent Evaluation Group (2017) Towards a clean world for all: an IEG evaluation of the World Bank Group's support for pollution management. World Bank, Washington, DC, unpaginated
9. World Health Organization (2016) Ambient air pollution: a global assessment of exposure and burden of disease. WHO, Geneva, 131p. apps.who.int/iris/bitstream/10665/250141/1/9789241511353
10. Demographia (2018) World urban areas (built up urban areas or world agglomerations). www.demographia.com/db-worldua.pdf
11. World Health Organization (2011) Guidelines for drinking-water quality, 4th edn. WHO, Geneva, 541p. apps.who.int/iris/bitstream/10665/44584/1/9789241548151
12. Tchoumwou PB, Yediou CG, Patiolla AK, Sutton DJ (2012) Heavy metals, toxicity and the environment. A review. EXS 101:133–164. www.ncbi.nlm.nih.gov/pmc/articles/PMC4144270/
13. Jaishankar M, Tseten T, Anbalagan N, Mathew BB, Beeregowda KN (2014) Toxicity, mechanism and health effects of some heavy metals. Interdiscip Toxicol 7:60–72. www.ncbi.nim.nih.gov/pmc/articles/PMC4427717/
14. Rizwan M, Ali S, Andrees M, Rizvi H, Zia-ur-Rehman M, Hannan F, Qayyum MF, Hadeez F, Ok YS (2016) Cadmium stress in rice: toxic effects, tolerance mechanisms, and management: a critical review. Environ Sci Pollut Res Int 23:17859–17879
15. Govind P, Madhurt S (2014) Heavy metals causing toxicity in animals and fishes. Res J Anim Vet Fish Sci 2:17–23
16. South China Morning Post (2013) High cadmium levels found in Guangzhou rice, South China. www.scmp.com/china/article/1240198/
17. Kobayashi I, Hagino N (1965) Strange osteomalacia by pollution from cadmium mining. Progress report WP 00359, Okayama University, pp 10–24
18. Kanu AS, Ashraf U, Bangura A, Yang DM, Ngaujah S, Tang X (2017) Cadmium (Cd) stress in rice; photo-availability, toxic effects and mitigation measures – a critical review. J Environ Sci Toxicol Food Technol 1:7–23. www.josr/journals.org
19. Chen H, Zhang W, Yang Y, Wang P, McGrath SP, Zhao F-J (2018) Effective method to reduce cadmium accumulation in rice grain. Chemosphere 207:699–707
20. Nordstrom DK (2002) Worldwide occurrence of arsenic in groundwater. Science 296:2143–2145
21. Pirrone N, Cinderella S, Feng X, Finkelman RB, Friedli HR, Leaner J, Mason R, Mukherjee AB, Stracher GB, Streets DG, Telmer K (2010) Global mercury emissions to the atmosphere from anthropogenic and natural sources. Atmos Chem Phys 10:5951–5964
22. Health Canada (2017) Mercury in fish. Consumption advice: making informed choices about fish. Health Canada, Ottawa, unpaginated

Chapter 7
An Example of Coastal Cities Hazard Exposure and Economics

Coastal cities' economics have developed in one or more than one sector depending on several factors. These include, but are not limited to, location and accessibility, climate, environment attractions for tourism and recreation, accessibility to organic and inorganic natural resources for internal use and export, or need to import resources. Populations in urban coastal centers continue to increase as global population grows and there is a great demographic change as rural citizens flock to cities. As previously noted, the 2018 global population of 7.6 billion citizens had 4.1 billion living in cities and 3.5 in rural areas. Projections indicate that in 2050, the global population will be 9.9 billion people. Cities will grow to 6.9 billion inhabitants from natural growth, from an influx of rural people, and in some cases from immigration. This leaves a population of 3 billion people in rural areas [1]. Rural citizens come to cities for employment opportunities, for better schools for their children, and for access to better healthcare. Coastal cities are absorbing much of the global population growth, mainly in Asia, Africa, and to some degree in South America.

7.1 City Populations and Assets Exposed to Coastal Flooding in 2007 and Projected to 2070

Table 7.1 shows population increases projected for 20 cities exposed to coastal flooding in 2007 and a projected population exposed to flooding in the 2070s [2, 3]. Coastal cities' assets can be financially enormous in terms of personal property, businesses, and governments. This is especially true where there are high populations and nationally important commercial interests and ports. The assets that are at risk today from global warming-driven floods and sea level rise are projected to

© The Author(s), under exclusive license to Springer Nature Switzerland AG 2020   63
F. R. Siegel, *Adaptations of Coastal Cities to Global Warming, Sea Level Rise,*
Climate Change and Endemic Hazards, SpringerBriefs in Environmental Science,
https://doi.org/10.1007/978-3-030-22669-5_7

Table 7.1 Top 20 cities ranked as most exposed to coastal flooding in the 2070s vs. 2007 populations from changes in climate and socio-economic status [2, 3]

Rank	Country	Urban agglomeration	2007 exposed population	2070s exposed population
1	India	Kolkata	1,929,000	14,014,000
2	India	Mumbai	2,787,000	11,418,000
3	Bangladesh	Dhaka	844,000	11,135,000
4	China	Guangzhou	2,718,000	10,333,000
5	Vietnam	Ho Chi Minh City	1,931,000	9,216,000
6	China	Shanghai	2,353,000	5,451,000
7	Thailand	Bangkok	907,000	5,138,000
8	Myanmar	Rangoon	510,000	4,965,000
9	The United States	Miami	2,003,000	4,795,000
10	Vietnam	Hai Phong	794,000	4,711,000
11	Egypt	Alexandria	1,330,000	4,375,000
12	China	Tianjin	956,000	3,790,000
13	Bangladesh	Khulna	441,000	3,641,000
14	China	Ningbo	299,000	3,305,000
15	Nigeria	Lagos	357,000	3,229,000
16	Cote D'Ivoire	Abidjan	519,000	3,110,000
17	The United States	New York, Newark	1,540,000	2,931,000
18	Bangladesh	Chittagong	255,000	2,865,000
19	Japan	Tokyo	1,110,000	2,521,000
20	Indonesia	Jakarta	511,000	2,248,000

15 in Asia, 3 in Africa, 2 in the United States
Note: Subsidence adds to problem in Ho Chi Minh City, Bangkok, Hai Phong, Jakarta

increase multifold in >50 years. Table 7.2 presents these figures for 20 cities for 2007 and the 2070s [2, 3]. Hence, people and assets should be protected by flood defenses. If proper defenses are not in place today, planning and investment to put them in place with utmost urgency is essential. This should include adaptation plans that can be instituted as reliable projections of future threats are recognized (e.g., continued sea level rise).

As populations exposed to flooding grow from 2007 to projected numbers in the 2070s, the exposure to flooding of assets value increases (Tables 7.1 and 7.2). Cities common to both are listed in Table 7.3 that shows the projected times increase for the nine cities. The increases are significantly greater in developing and less developed economies than for those in developed ones.

Table 7.2 Top 20 cities ranked as to assets exposed to coastal flooding in the 2070s vs. assets exposed in 2007 from changes in climate and socio-economic status [2, 3]

Rank	Country	Urban agglomeration	2007 exposed assets (US$Billions)	2070s exposed assets ($Billions)
1	The United States	Miami	416.29	3513.00[a]
2	China	Guangzhou	84.17	3357.72
3	The United States	New York–Newark	320.20	2147.35
4	India	Kolkata	31.99	1961.44
5	China	Shanghai	72.86	1771.17
6	India	Mumbai	46.20	1598.05
7	China	Tianjin	29.62	1231.48
8	Japan	Tokyo	174.29	1207.07
9	China	Hong Kong	35.94	1163.89
10	Thailand	Bangkok	38.72	1117.54
11	China	Ningbo	9.26	1073.93
12	The United States	New Orleans	233.69	1013.45
13	Japan	Osaka–Kobe	215.62	968.96
14	The Netherlands	Amsterdam	128.33	843.70
15	The Netherlands	Rotterdam	114.89	825.68
16	Vietnam	Ho Chi Minh City	26.86	652.82
17	Japan	Nagoya	109.22	623.42
18	China	Qingdao	2.72	601.59
19	The United States	Virginia Beach	84.64	581.69
20	Egypt	Alexandria	28.46	563.28

13 in Asia, 4 in the United States, 2 in Europe, 1 in Africa
[a]A 2018 report estimated that by 2045, flooding exacerbated by sea level rise would threaten 12,095 homes valued at US$6.4 billion, an asset far higher than the US$3.5+ billion estimated above for 2070 [4]

7.2   Projected Impact of Coastal Flooding on Cities' GDPs with a 7.9 in. (20 cm) Sea Level Rise

Economists may review financial figures in terms of how an event such as flooding in a coastal city and losses therefrom affects the gross domestic product (GDP). The GDP is expressed as a percentage. It is often defined as the sum of all goods and services produced by a city (or nation). GDP is more specifically defined by the equation

Table 7.3 A review of the data of nine coastal cities common to Tables 7.1 and 7.2 shows that as a population exposed to flooding increases from 2007 to one projected for the 2070s, there are projected increases as well to exposure to flooding asset value, but to a greater degree

City	Times increase in exposed population 2007–2070	Times increase in exposed assets 2007–2070
Ningbo	11	115.9
Kolkata	7.26	61.3
Tianjin	4	41.5
Mumbai	4.1	34.5
Ho Chi Minh City	4.8	24.3
Alexandria	3.3	19.8
Miami	2.4	8.4
Tokyo	2.2	6.9
New York–Newark	2	6.7

6 in Asia, 2 in the United States, 1 in Africa

Table 7.4 Coastal cities that may face the highest financial average annual loss as a percentage of their GDP by 2050 from socio-economic changes, subsidence, sea level rise of 20 cm (7.9 in.), and adaptation to maintain at least a 100-year exposure defenses against coastal flooding [5]

City	Average annual loss US$ millions	% Loss of GDP
1. Guangzhou	13,200	1.46
2. Mumbai	6,414	0.49
3. Kolkata	3,350	0.36
4. Guayaquil	3,189	1.08
5. Shenzen	3,136	0.40
6. Miami	2,549	0.36
7. Tianjin	2,276	0.30
8. New York–Newark	2,056	0.08
9. Ho Chi Minh City	1,953	0.83
10. New Orleans	1,864	1.42
11. Jakarta	1,750	0.22
12. Abidjan	1,023	0.89
13. Chennai	939	0.14
14. Surat	928	0.26
15. Zhangjiang	891	0.55
16. Tampa–St. Petersburg	859	0.29
17. Boston	793	0.14
18. Bangkok	734	0.09
19. Xiamen	724	0.29
20. Nagoya	644	0.30

13 in Asia (4 in India) (5 China) (1 Japan) (3 SE Asia)
5 in the United States—1 in South America, 1 in Africa

$$GDP = C + I + G + \left(Ex - Im\right)$$

C = spending by consumers
I = total investment (spending for goods and services by businesses)
G = total spending by government (federal, state, local)
$(Ex - Im)$ = net exports (exports − imports)

A reasonable growth of GDP is 2–3%. A falling GDP over a period of time will alert monitors of global economics to the possibility of a coming recession. Financial institutions can then adapt by planning and implementing actions to damp or eliminate the possibility of a recession that might otherwise become a depression. In the situation of major coastal cities that contribute significantly to a nation's GDP, necessary investment in protection against natural hazards, such as the example of flooding presented in this section, is warranted but the financial implications of doing what is deemed necessary is up for discussion. Table 7.4. shows the negative effects on GDP in coastal cities exposed to flooding for some cities that might not make the investments to upgrade flood defenses with the purpose of minimizing a city's exposure to impacts on citizens and a city's assets. As measured by GDP loss, Guangzhou in Asia, New Orleans in North America, Guayaquil in South America, Abidjan in Africa, and Ho Chi Minh City in Southeast Asia would suffer most economically [5].

Basically, there are two options that decision-makers can follow, one of which gives momentary relief at great expense and the other of which requires substantial investments to preserve much more valuable assets and of course reduces death and injury to urban citizenry. The first option is to suffer through a hazard and its disastrous impacts and then spend huge economic resources (if they are available) to **repair, restore, and rebuild** what was damaged or destroyed and also suffer through burying the dead and caring for the injured. The other is to make major capital investments before a hazard impacts in order to **protect** citizens and assets by **preventing** to a good degree hazards impacts and thus **preserve** a city and its inhabitants.

7.2.1 2200 Global GDP Loss?

A far out evaluation to 2200 of the economic cost of coastal flooding caused by human-driven climate change compared to 2016 was generated using a dynamic computer model that accounted for projected sea level rise, migration, trade, and technological advances during the twenty-first century [6, 7]. The sea level rises used in the study were the Intergovernmental Panel on Climate Change estimations for 2000–2100 at the moderate Representative Concentration Pathway (RCP 4.5) range of 0.4–0.9 m (16–36 in.) [8] and a probabilistic moderate RCP 4.5 projection for 2100–2200 of 1.25 m (50 in.) [7].

Coastal cities will take a direct hit and permanent inundation will require the costly adaptation to move economic activities inland. This may be possible for the United States but not so for example, for the low-lying Netherlands or island nations. It will cause population migration inland or to other countries and will affect economies because of changes in the movement of goods and delivery of services internationally. The displaced population by 2200 will be 1.46% of the world total vs. a 0.58% displaced population of the world total by 2100. The worst national projection for 2200 was for Vietnam that the computer data indicated could lose 40.5% of its land area, displace 21.8% of its population, and suffer a 21.5% loss of real GDP. The physical locations of world economy centers would change as there are relocations from land permanently lost to inundation to safe land. This will result in new nodes of economic development. The data input to the dynamic computer model used indicates that economic adaptation of investment and migration to counter the effects of permanent coastal flooding, or lack thereof, could affect a real loss in global GDP for 2200 in two ways. With adaptation the loss is predicted to be 0.11%. Without the adaptation investment the calculated real loss of global GDP for 2200 would increase to 4.5% [6].

This is an interesting exercise based on data projections for sea level rise, migration, trade, and technological innovation to 2200. However, uncertainties clearly exist in the basics used in predicting future global environmental, economic, and demographic changes. Today, the rate of average sea level rise is ~3.3 mm annually, but this will likely change to higher values because of increases in global warming that affect melting of alpine glaciers, ice caps, and ice shelves worldwide and heat expansion of the oceans thereby increasing their volumes. Then there is world population growth and the numbers that will inhabit coastal locations. Will global population stabilize at 11.2 billion people in 2100? What number can the earth sustain with water, food, and other necessities and amenities that define a satisfactory quality of life? Computer assessments such as discussed above, based on available real data and data projections, are at best educated estimations. Nonetheless, they demonstrate that calculations of future loss models should include dynamic changes such as those cited in this section.

References

1. World Population Data Sheet (2018) Population Reference Bureau, Washington, DC
2. Nicholls RJ, Hanson S, Herweijer C, Patmore N, Hallegatte S, Corfee-Moriot J, Chateau J, Muir-Wood R (2007) Ranking of the world's cities most exposed to coastal flooding today and in the future. Organization for Economic Cooperation and Development, Paris, 10 pp. https://idrc.info/fileadmin/user_upload/idrc/former…
3. Hanson S, Nicholls R, Ranger N, Hallegatte S, Corfee-Morlot J, Herweijer C, Chateau J (2011) A global ranking of port cities with high exposure to climate extremes. Climatic Change 104:89–111. https://link.springer.com/content/pdf/10.1007/s10584-010…
4. Union of Concerned Scientists (2018) New study finds 1 million Florida homes worth $351 billion will be at risk from tidal flooding. Washington, DC, 28 p

5. Hallegatte S, Green C, Nicholls RJ, Morlot JC (2013) Future flood losses in major coastal cities. Nat Climate Change 3(9):802–806
6. Desmet K, Kopp RE, Kulp SA, Kristian, Oppenheimer M, Hansberg ER Strauss BH (2018) Evaluating the economic cost of coastal flooding. National Bureau of Economic Research, Working Paper No. 24918, Cambridge, MA, 34 p. www.princeton.edu/erossi/EECCF.pdf
7. Kopp RE, Horton RM, Little CM, Mitrovica JX, Oppenheimer M, Rasmussen DJ, Strauss RH, Tebaldi C (2014) Probabilistic 21st and 22nd century sea-level projections at a global network of tidal gauge sites. Earth's Future 2:383–406. American Geophysical Union, Washington, DC. https://agupubs.onlinelibrary.wiley.com/doi/10.1002/2014EF000239
8. International Panel on Climate Change (2013) Fifth assessment report. In: Climate change 2013. The physical science basis. Summary for policy makers. Cambridge Univ. Press, Cambridge, 29 p. http://www.ipcc.ch/report/ar5/wgl/

Chapter 8
Decisions, Costs, Funding to Protect Coastal Cities: Populations and Assets (Personal and Municipal/National)

8.1 Investment Decisions for Coastal Cities Protection

Decisions on where and what investments should be made to build defenses to protect people and assets in coastal cities from exposure to hazard events are generally based initially on a **cost-benefit analysis**. Secondly, decisions ideally consider which investment or investments will most benefit a city's general citizenry and not an economically advantaged or politically influential segment of a population. A third consideration in decision-making is how much municipal/national funding is initially available and how much funding can be obtained by low cost loans from institutions such as the World Bank and regional development banks that serve most areas (e.g., the European Bank for Reconstruction and Development, the Asian Infrastructure Investment Bank, the African Development Bank, and the Inter-American Development Bank). In addition, many national agencies in developed countries award grants and offer technical assistance to less-developed and developing countries to support well-planned development projects. These can originate, for example, from the United States, the European Union, France, Germany, Japan, Norway, the Netherlands, the United Kingdom, and others. The United Nations Development Program can provide technical assistance in lieu of funding.

8.2 Cost Determinants

Costs for a coastal defense project will vary significantly from country to country or within a country (See Sect. 3.1.1) Initial costs would be for planning design and engineering assessment. This will vary with the degree of coastal city urbanization with higher costs for non-uniform physical and demographic conditions. Added to this are material costs (e.g., concrete, stone, sand) that will vary with accessibility

F. R. Siegel, *Adaptations of Coastal Cities to Global Warming, Sea Level Rise, Climate Change and Endemic Hazards*, SpringerBriefs in Environmental Science, https://doi.org/10.1007/978-3-030-22669-5_8

(nearby, transported in, or even imported). Labor costs will vary among countries and within a country itself, a factor that affects whether manual labor or machines are used in a project. Land purchase costs can be expensive in coastal cities where land is scarce and this can influence the type of defense system that is selected for installation. Lastly, there is the cost of management during and after a project is completed and maintenance costs that are established after yearly inspections of defense structures. Management also monitors conditions to observe changes (e.g., sea level rise, drainage basin construction) that may require adaptation of a defense system to better protect coastal city inhabitants and assets [1].

The cost for a specific defense system for different reaches of a coastal city depends on the length of a preferred option for each stretch and its width and height dimensions. To this is added whatever changes have to be implemented with respect to its alignment with recurring direction of incoming events and/or sea level rise. These factors plus those given in the previous paragraph allow a calculation of a cost price per meter or foot of installation and hence a total project cost [1, 2].

Costs to adapt coastal defenses to sea level rise and other drivers of floods differ among countries as shown for dike/levee heightening with estimates from the Netherlands, New Orleans, and Vietnam. Dike heightening in the Netherlands of 0.8, 1.8, and 2.4 m for 1 km at three locations averaged M€4.41, M€6.095, and M€7.79, respectively. In New Orleans, heightening a dike 1 m for 1 km ranged from M€5 to 8. The estimated cost to heighten a dike 1 m for 1 km in Vietnam ranged from M€0.7 to 1.2. How high a dike should be to protect coastal cities from flooding depends on the recorded history of past flood events and/or geologic studies that can reveal these events preserved in the rocks underlying the coastal regime. The Netherlands builds to protect its coast from a once in a 4000- or a 10,000-year event. New Orleans builds to protect the city from a 100 year event. The Vietnam standard design is to build against a 20–25 year recurrence of flooding to protect people and assets. Given Vietnam's low-lying areas, 70% of the population is subject to risk of damaging flooding. Recent studies on risk-based design of coastal flood defenses show that at a minimum the Vietnam defenses should be heightened to withstand at least a 50-year recurrence interval. Nonetheless, to better protect citizens and ongoing economic development projects, the defenses should be able to withstand at least a statistical 100-year flood recurrence interval [3, 4].

8.3   Differences in a Nation's Management, Monitoring, and Maintenance Costs for Coastal Cities Defenses

An example of the managing, monitoring, and maintenance costs for coastal protection against flooding is that of the Netherlands. In 2000, the Netherlands spent €360 million annually to monitor and maintain 3600 km of defenses annually or ~€100,000/km/year. The defenses include dikes and levees, dams, sluices, and sea gates. It should be noted that the Netherlands presents a special situation because 20% of the country is below sea level and 38% is below high tide level. In 2016, the

Dutch expenditure for defenses maintenance more than tripled to €1.2 billion. This illustrates that a projected changing cost factor has to be included in budgetary consideration to assure continuity of the monitoring and maintenance oversight. Without proper maintenance, flood defenses may lose effectiveness after 50 years. In contrast, the expenditure by Vietnam to maintain its 2000 km dike system is €40 million annually or €20,000–€30,000/km. The large difference is due in part to the design of dikes for the level of flood protection determined by a national government to be necessary with respect to recurring flood events as described in the previous paragraph.

References

1. Hillen MM, Jonkman SN, Kanning W, Kok M, Geldenhuys MA, Stive MJF (2010) Coastal defence cost estimates. Case study of the Netherlands, New Orleans and Vietnam, Communications on Hydraulic and Geotechnical Engineering, Delft Technical University, 87 p
2. Jonkman S, Hillen N, Marten M, Nicholls RJ, Kanning W, van Ledden M (2013) Costs of adapting coastal defenses to sea level rise—new estimates and their implicitons. J Coastal Res 29:1212–1226
3. Mai CV, van Gelder PHAJM, Vrijling JK, Mai TC (2008) Risk analysis of coastal flood defences—a Vietnam case. In: 4th International Symposium on Flood Defence, Toronto, Canada, 8 p
4. Mai CV, van Gelder PHAJM, Vrijling JK (2010) Risk based design of coastal flood defences—a Vietnam case. In: Reliability, risk and safety: theory and applications. Taylor & Francis, London, pp 1125–1132

Epilogue

Physical, chemical, and biological technology and engineering expertise are available now to design and/or adapt defenses that can mitigate the impacts of global warming, climate changes, and other hazards that present threats to many coastal cities. It is now that defenses should be planned and put in place with urgency where most needed.

Funding is a problem, but with prioritized national funds for coastal defenses, controlled lending from international development banks and grants from countries with support agencies and those with substantial foreign reserves, important progress in planning defenses, and their subsequent construction or application can be made fairly rapidly. Investment now will preserve a coastal city's economic base and, hence its productivity and its contribution to a national GDP, especially in less developed and developing countries.

This is what peoples of the world and their governments should be moving towards sooner rather than later. Delay means decay and needless dangers to coastal communities whether mega-cities or other high population cities or those with smaller populations but with economically important seaports and/or other attractions. If this book stimulates responses in any of the at risk coastal communities that create or improve their defenses against existing and future global warming and climate change problems, or other natural and anthropogenic hazards, it was worth writing.

Appendix: List of Coastal Cities with a One Million or More Population Potentially at Risk from Global Warming, Climate Change Events, and Natural Hazards [1]

Urban area	Urban population
Tokyo, Japan	38,050,000
Jakarta, Indonesia	32,275,000
Manila, Philippines	27,280,000
Seoul Incheon, South Korea	24,210,000
Shanghai, China	24,115,000
Mumbai, India	23,265,000
New York Environs, USA	21,575,000
Sao Paulo, Brazil	21,100,000
Guangzhou, China	19,965,000
Dhaka, Bangladesh	17,425,000
Osaka–Kobe–Kyoto, Japan	17,165,000
Bangkok, Thailand	15,975,000
Los Angeles Environs, USA	15,620,000
Buenos Aires, Argentina	15,520,000
Istanbul, Turkey	13,995,000
Lagos, Nigeria	13,910,000
Karachi, Pakistan	13,255,000
Rio de Janeiro, Brazil	11,990,000
Lima, Peru	11,355,000
Ho Chi Minh City, Vietnam	10,690,000
London, UK	10,585,000
Chennai, India	10,555,000
Nagoya, Japan	10,105,000
Taipei, Taiwan	8,605,000
Wuhan, China	7,980,000
Luanda, Angola	7,560,000
Hong Kong, China	7,380,000

© The Author(s), under exclusive license to Springer Nature Switzerland AG 2020
F. R. Siegel, *Adaptations of Coastal Cities to Global Warming, Sea Level Rise, Climate Change and Endemic Hazards*, SpringerBriefs in Environmental Science, https://doi.org/10.1007/978-3-030-22669-5

Urban area	Urban population
Boston–New England Environs, USA	7,315,000
Quanzhou, China	6,720,000
San Francisco–San Jose, USA	6,540,000
Nanjing, China	6,525,000
Surat, India	6,200,000
Miami, USA	6,195,000
Suzhou, China	6,175,000
Qingdao, China	5,930,000
Singapore, Singapore	5,930,000
Philadelphia Environs, USA	5,575,000
Yangon, Myanmar	5,550,000
Fuzhou, China	5,400,000
St. Petersburg, Russia	5,175,000
Abidjan, Ivory Coast	5,145,000
Dar es Salaam, Tanzania	4,980,000
Alexandria, Egypt	4,960,000
Kuwait, Kuwait	4,860,000
Barcelona, Spain	4,840,000
Dalian, China	4,600,000
Wenzhou, China	4,460,000
Casablanca, Morocco	4,410,000
Sydney, Australia	4,390,000
Surabaya, Indonesia	4,325,000
Melbourne, Australia	4,305,000
Xiamen, China	4,265,000
Accra, Ghana	4,260,000
Colombo, Sri Lanka	4,205,000
Dubai, UAE	3,990,000
Cape Town, South Africa	3,980,000
Jeddah, Saudi Arabia	3,975,000
Seattle, USA	3,860,000
Santo Domingo, Dominican Republic	3,815,000
Algiers, Algeria	3,780,000
Changzhou, China	3,770,000
Ningbo, China	3,735,000
Naples, Italy	3,695,000
Zhangjiagang, China	3,595,000
Zhongshan, China	3,575,000
Durban, South Africa	3,515,000
Porto Alegre, Brazil	3,485,000
Athens, Greece	3,470,000
Tel Aviv, Israel	3,465,000
Dakar, Senegal	3,450,000
Recife, Brazil	3,425,000

Urban area	Urban population
Chittagong, Bangladesh	3,400,000
Douala, Cameroon	3,310,000
Izmir, Turkey	3,280,000
San Diego, USA	3,255,000
Busan, South Korea	3,225,000
Nanchang, China	2,935,000
Guayaquil, Ecuador	2,890,000
Caracas, Venezuela	2,890,000
Pyongyang, North Korea	2,885,000
Baku, Azerbaijan	2,865,000
Port-au-Prince, Haiti	2,850,000
Maputo, Mozambique	2,735,000
Tampa Environs, USA	2,720,000
Lisbon, Portugal	2,715,000
Rotterdam–The Hague, The Netherlands	2,675,000
Fukuoka, Japan	2,635,000
Cebu, Philippines	2,635,000
Taichung, Taiwan	2,596,000
Tangshan, China	2,590,000
Mogadishu, Somalia	2,590,000
Kochi, India	2,585,000
Kaohsiung, Taiwan	2,555,000
Cali, Colombia	2,540,000
Shantou, China	2,515,000
Vancouver, Canada	2,335,000
Baltimore, USA	2,335,000
Yantai, China	2,310,000
Beirut, Lebanon	2,285,000
Tunis, Tunisia	2,280,000
Haikou, China	2,250,000
San Juan, Puerto Rico	2,135,000
Maracaibo, Venezuela	2,130,000
Brisbane, Australia	2,120,000
Hamburg, Germany	2,115,000
Portland, USA	2,075,000
San Antonio, USA	2,070,000
Belem, Brazil	2,060,000
Port Harcourt, Nigeria	2,060,000
Kunshan, China	2,055,000
Brazzaville, Congo	2,045,000
Huizhou, China	2,025,000
Havana, Cuba	2,020,000
Perth, Australia	1,945,000

Urban area	Urban population
Visakhapatnam, India	1,905,000
Sacramento, USA	1,885,000
Lome, Togo	1,845,000
Abu Dhabi, UAE	1,840,000
Barranquilla, Colombia	1,810,000
Johor Bahru, Indonesia	1,810,000
Conakry, Guinea	1,785,000
Freetown, Sierra Leone	1,755,000
Zhangzhou, China	1,750,000
Doha, Qatar	1,745,000
Kitakyushu, Japan	1,740,000
Taizhou, China	1,725,000
Santos, Brazil	1,700,000
Semarang, Indonesia	1,690,000
Yingkou, China	1,680,000
Wuhu, China	1,675,000
Yangzhou, China	1,665,000
Zhuhai, China	1,660,000
Amsterdam, The Netherlands	1,660,000
Novosibirsk, Russia	1,660,000
Marseille, France	1,630,000
Panama City, Panama	1,595,000
Valencia, Spain	1,595,000
Makassar, Indonesia	1,590,000
Stockholm, Sweden	1,565,000
Al Manamah, Bahrain	1,555,000
Monrovia, Liberia	1,545,000
San Salvador, El Salvador	1,540,000
Valencia, Venezuela	1,540,000
Changshu, China	1,535,000
Auckland, New Zealand	1,530,000
Virginia Beach–Norfolk, USA	1,495,000
Davao City, Philippines	1,490,000
Porto, Portugal	1,490,000
Jiangmen, China	1,460,000
Penang, Malaysia	1,460,000
Batam, Indonesia	1,415,000
Mersin, Turkey	1,410,000
Hiroshima, Japan	1,390,000
Sendai, Japan	1,365,000
Montevideo, Uruguay	1,320,000
Nantong, China	1,310,000
Copenhagen, Denmark	1,305,000
Qinhuangdao, China	1,300,000

Urban area	Urban population
Odessa, Ukraine	1,100,000
Bucaramanga, Colombia	1,090,000
Basra, Iraq	1,085,000
Cartagena, Colombia	1,060,000
Pointe Noire, Congo	1,045,000
Wahran, Algeria	1,040,000
Da Nang, Vietnam	1,040,000
Maceio, Brazil	1,025,000
Oslo, Norway	1,025,000
Antwerp, Belgium	1,020,000

Index